Funding High-Tech Ventures

Richard L. Manweller

The Oasis Press
Grants Pass, Oregon

Published by The Oasis Press®/PSI Research

Editor: Karen Billipp
Interior design by Eliot House Productions
Cover design by Steven Burns
Cover photography by Rob Rebman

Please direct any comments, questions, or suggestions regarding this book to:
The Oasis Press®/PSI Research:

Editorial Department
300 North Valley Drive
Grants Pass, OR 97526
(541) 479-9464
(541) 476-1479 *fax*
psi2@magick.net *e-mail*

The Oasis Press® is a Registered Trademark of Publishing Services, Inc., an Oregon corporation doing business as PSI Research.

Library of Congress Cataloging-in-Publication Data
Manweller, Richard L., 1936-
Funding high tech ventures / Richard L. Manweller. -- 1st ed.
p. cm. -- (PSI successful business library)
ISBN 1-55571-405-6 (paper)
1. Venture capital. 2. High technology industries -- Finance.
I. Title. II. Series.
HG4751.M36 1997
658.15'22--dc21 97-35132
CIP

Printed and bound in the United States of America
First Edition 10 9 8 7 6 5 4 3 2 1 0
Printed on recycled paper when available

Contents

Preface . . . vii
Before You Start . . . x

I **Company Prerequisites . . . 1**
Company Standing . . . 1
Personnel Profile: The Right Stuff . . . 3
The Objective . . . 3

II **Overview of Funding Sources: The Mortgagors and Partners . . 4**
Banks: The Sunny Day Umbrella Lenders . . . 4
Venture Capitalists: The Companynappers . . . 5
Synergistic Partners . . . 5
Informal Equity Investors: The Angels . . . 6
Research and Development Ventures: Asset Sharing . . . 7
What to Avoid . . . 8

III **Planning: The Groundwork . . . 9**
Overview of Business Plan and Appendix Documentation . . . 9
Who Does What? . . . 11
Management Leader Tasks . . . 12
Technology Leader Tasks . . . 22
Marketing Leader Tasks . . . 23

IV **The Business Plan Format . . . 25**

Putting It All Together . . . 25

The Appendices Format . . . 27

V **Finding Investors: The Quest . . . 28**

Sources . . . 28

VI **Presentation: The Technique of Selling Carrot Seeds . . . 31**

Strategy . . . 31

Format . . . 33

VII **Confidentiality and Nondisclosures . . . 34**

Nondisclosure Agreement . . . 34

Title Page Nondisclosure Statement . . . 36

VIII **Agreement: The Share-Cropping Terms . . . 37**

Important Reminders . . . 37

Proposal . . . 38

Securities . . . 39

Apportionment . . . 40

Control . . . 41

Finance Phasing . . . 41

IX **Now That You've Got It . . . Planting Straight Rows . . . 43**

To Spend or Not to Spend . . . 43

To Reveal or Not to Reveal . . . 44

X **Help 46**

XI **Addendum . . . 49**

Business Plan . . . 51

- Title Page . . . 51
- Executive Summary . . . 53
- Management . . . 59
- Marketing . . . 62
- Product . . . 67
- Product and Marketing Positioning . . . 75
- Sales . . . 83

Operations 88
Financial 91
Financial Spreadsheets 95
Glossary 114

Worksheets 121
Sales Forecast Chart. 123
Summary Chart 124
Return on Investment (ROI) Analysis. 125
Projected Income Statement 126
Cash Flow Statement 132
Balance Sheet 138

Index 145

Preface

"Do nothing, lest we fail?"

A Case History

The Start-Up

When my partner and I first went in search of funding, we envisioned we could show up at any investor's door and be welcomed with wide open, hungry arms. Our unbelievably sophisticated, state-of-the-art, high-tech "mousetrap" would so overwhelm the investors, that selecting the winner, out of all the investor candidates, would be our only problem.

Perhaps we were sleepwalking, because no one answered the door. Neither partner could believe it; no candidates at all — what a letdown.

Our company started its recovery with the above Lincolnesque quote — it's now permanently located on the wall above my desk. We decided that we would rather not fail nor quit; but failing would be better than quitting. I hit the book stores and libraries, and read every how-to book available on start-ups and business planning. As a result, we developed an unbelievably sophisticated, state-of-the-art business plan and . . . got the same results. Why?

That's what we asked the investor firms — why? Investors replied: "We don't fund your type of future company"; "We aren't funding start-ups at this time"; "What we don't understand, we don't fund"; "We only fund companies located in our area"; and a variety of other canned responses.

We reread the quote and repeated it to our frowning spouses.

In the end, we did get funding — more than we expected. We were all elated, especially the spouses. How did we do it?

After rereading the books, I deduced that they had taught us how to create an MBA business plan; not a salable, investor friendly plan. The how to select the right type of investor; the how to prepare for and present a business proposal to investors; the how to get the investor to act, and the how to get an equitable arrangement and agreement with investors were completely missing. What they all lacked, was in-depth investor strategy. Strategy that provided more than "try this."

We rewrote the business plan, this time directing it to the intended reader — the investor. On our third investor meeting, we hit pay dirt.

Don't go through the same, although finally successful, trial-and-error process that we did to get our first company funded. Don't knock on doors that won't open. If the doors do open, don't present plans that have no chance of getting funded. Business plans are sales documents. Learn how to create one by reading this book.

The Emerging and Established Company

Our start-up successfully completed its product development and testing and trade show presentations, and was ready to initiate sales. The company was quickly metamorphosing from an engineering research and development group into a marketing and sales company. To attack the marketplace, the company now needed additional funding to expand production, marketing, and sales facilities. What to do?

Starting with our original business plan, we greatly reduced the technical sections and expanded on the operations, marketing, and sales sections. The company was ready to mass produce and sell the product and that's what we emphasized. Using the same investor strategy utilized earlier and reiterated throughout this book, we successfully obtained the funding required to take the company to profitability.

And our spouses felicitously said they knew we could do it.

❖ ❖ ❖

Whether you're striving to fund a start-up, an emerging company, or an established company, *Funding High-Tech Ventures* will provide you with the necessary methodologies and winning strategies to gain an unfair advantage over the multitudes. This is one race that will always go to the smart and knowledgeable, with the strong and swift second and the uninformed left at the starting gate. Of course, the strong, swift, and knowledgeable have a really inequitable advantage.

This is a book about investors and what you have to do to obtain their investment capital. The competition for these funds is enormous. Take the unfair advantage and absorb this book!

If you like what you read, you can thank that foreign minister, the one that stated American managers are lazy and incompetent. At that time, I was working over 80 hours a week putting a new company together and I failed to see the humor in his remark. I suppose his real meaning got lost in the translation.

American managers are not lazy, nor are they incompetent. I do believe, though, that our highly competitive, free enterprise system, has created managers who have become habituated to not revealing anything to anybody. American businesspeople preserve their acquired knowledge, like chefs withhold their recipes' ingredients. As a result, every new entrepreneur has been and is still "re-inventing the wheel." Thus, the book.

May the luck of the work hard and smart go with you!

Before You Start

Purpose

The primary purpose of *Funding High-Tech Ventures* is to provide you with the knowledge needed to obtain quality funding for your business. This workbook will take you step-by-step through the following:

- Determining your present company's status;
- Determining the type of funding best for your company;
- Planning, designing, and completing an investor friendly, high-tech business plan, with a complete business plan included for reference, and worksheets;
- Finding quality investors;
- Preparing and presenting your business proposal and plan to investors;
- Preparing and negotiating the funding agreement; and
- Managing the funded enterprise.

I

Company Prerequisites

Company Standing

This book will take you through the steps necessary to open the doors to financial benefactors, assuming your company and its personnel are properly positioned. Unbelievable luck aside, one can sell carrots or even carrot seeds, but; not visions of carrots. If you can answer the following questions affirmatively, your company is at the necessary state of emergence to begin the quest for funding.

1. Can the product or service be demonstrated?

A product or service idea, not yet operable or proven, has little chance of obtaining funding unless the members of your management team have a magnificent track record or are known magicians or both. Ideally, a prototype is available for demonstration or at the very least a simulation of the product has been developed to demonstrate its future capabilities and performance. A possible exception is an idea, backed up with thorough research, which is such an obvious winner that investors believe it simply can't fail. Here again, the determining factor will be your management teams' and company's previous history of accomplishments.

Investors will require that the technology is available today, that it has unique benefits, has no legal restrictions, is easily maintainable, and fits with existing or planned production capabilities. It's difficult to determine if your product conforms to these requirements without a working model.

Without an available prototype, investors see an intolerable amount of risk from unknowns. They fear that competitors could easily duplicate the product in a short time. The end result is either no funding or funding at an excessive loss of ownership.

Balance the company's present economic condition and the cost of self-funding the prototype development with your future wants and your capability to sell dreams. One halfway measure that has worked is to develop an impressive software simulation of the future product. Caution: it may take as long to find a dream funder as it takes to create the prototype.

2. Are marketing studies complete?

Investors want evidence that the management team has a clear idea of who will purchase the product or service and why. A nonmarket-driven product will not get funded. If you don't know if you have such a product, now is the time to find out. Market driven does not imply that the product is not unique; but it does say that you believe, and have supporting evidence, that lots of revenue producers will want and need this product.

Don't ignore gut feelings, but never believe it's enough!

3. Is the basic management team consisting of a management leader, a technical leader, and preferably including a marketing leader, complete and ready?

The management leader needs to know where the company is going and how it plans to get there. The technical leader needs to thoroughly understand the product. An additional leader that knows how to find product buyers is a plus. In the eyes of investors, management teams are much more important than products.

4. Are we ready and able?

If not, determine what needs to be modified or added to bring the company up to a pre-funding status. Outside services can help in some cases, especially in the market analysis area. If the problem is the product, the management team must determine the acceptable level of completion needed to excite and entice investors. Conversely, do not analyze the product to destruction, as many large companies do; but, generally, you only get one shot at investors — it's best to be completely prepared and ready.

Neither should you fall victim to the ready-aim-aim-aim syndrome — be willing to fire!

Personnel Profile: The Right Stuff

If the present management team leader lacks great instincts, persistence, determination, resolve, confidence, tenacity, stability or belief in the company and product, you better search now for another captain. Investors want reassurance that the team leaders not only have the skills, but the true grit necessary to carry the company through the inevitable crises. In addition, the quest for funding will require that the management leader exhibit these attributes plus flexibility, some stubbornness and an overwhelming belief that funding will come. Losing is not a conceivable outcome to real entrepreneurs.

The management leader must exude positiveness. Negative thinking begets negative results and you don't have the time to wallow around in a quagmire. If you want a winner, believe in it and promote it!

Persistence, determination, resolve, confidence, tenacity, stability and belief — worth recalling when those pesky bugs congregate and become behemoths.

The Objective

Get the funding!

The goal is not to create love-me-or-leave-me egos, as investors surely will leave you. Nor should you be proposing to build super-duper-just-beyond-the-state-of-the-art products that you feel the investors can't live without, as they surely can! And if you covet your neighbor's Porsche, keep it to yourself. At this point, the goal is not developing, building, or selling the product or developing your self-wealth or position in life. When you believe you may have finally found the investor, in the heat of the battle over ownership rights, don't forget the objective. Get the Funding! Your spouse can help here with a reminder — 100 percent of nothing feedeth not. A later chapter covers how not to end up with 00.0 percent of everything where you eateth leftovers.

II

Overview of Funding Sources: The Mortgagors and Partners

Banks: The Sunny Day Umbrella Lenders

If your company is an emerging company, don't waste your time. Useful support from a bank is far down the road, and occurs when they want you, not vice-versa. By all means, though, develop a good banking relationship; it's imperative that you do. Your future growth may well depend on obtaining good financial references from your bank. Certainly, if you attempt to purchase parts or equipment from venders on non-COD terms, a bank report on your financial condition will be required.

Existing companies should have, by now, established a credit line, which they can draw upon when needed. Funding product developments this way or by any other type of bank borrowing is very expensive and detrimental to a company's short-term financial sheets. The temptation to use bank financing to fund product developments is often driven by the wish to retain company stock ownership. Additionally, in down periods when profits can't support the development of needed new products, bank financing may be the only option. The Synergistic Partner or R&D Venture methods described below are options worth investigating before pursuing bank financing.

Venture Capitalists: The Companynappers

The term Venture Capitalists (V/C) is an oxymoron, it should be U/Bs (Unadventurous Brokers), especially in hard times. V/Cs today prefer to invest in products which are being developed by sedate, well entrenched companies. If that's your company, V/Cs are a good source to approach for additional equity funding. For emerging companies, there's a slight chance, with a lot of luck, that there exists out there an elusive V/C who would invest in your company. The amount of time, energy, and money you would have to spend to find the right one may be intolerable.

Some V/Cs are extremely knowledgeable and helpful, but many aren't. In addition, because the competition for their time and funds is enormous, some have developed personality traits more akin to professional wrestlers than professional investors.

If you've got the time, try it. You'll get a real education in how to string along future vendors.

Synergistic Partners

"Synergy: Having the capacity to interact such that the total effect is greater than the sum of the individual effects." —Webster's Dictionary

"Partners, with independent assets, which, by combining these assets, will provide each other with economical, accelerated benefits." —Moore

The merger of synergistic corporate partners to develop your company or product or both is an excellent means of obtaining quality financing. This combination of complimentary assets can benefit both entities and be mutually rewarding.

The Partners' Assets

Your company's value will consist of tangible product assets and intangible assets — copyrights, trademarks, licenses, patents, the experience and proprietary knowledge of your personnel, start-up losses, and other good will. The value of good will is at least equal to the total capital a competing company would have to expend to get to where your company is today.

The investing company's most valuable asset, for your company, is its available funding and financial stability. Other nearly as important assets consist of resources such as: facilities, accounting and legal services, manufacturing and service capabilities, marketing and intangible assets. Intangible assets like vendor, bank, community, state, country, and international name recognition, as well as in-depth business knowledge can be of considerable benefit to unrecognized start-ups.

Combined Assets

A winning proposition for both the company and investing partner. For the company, the investing partner's resource and funding benefits are obvious. In addition, name dropping the investing partner's name, when needed, can be invaluable in negotiations with vendors and banks. It can also help in signing up customers to test the prototype product (alpha-site tests) and the finished product (beta-site tests), and open the door to other strategic partners, community and national leaders, and future customers. The investing partner may also be a beta-site and first customer buyer, providing early evidence of customer acceptance of the product. Additionally, established companies, because of their early-on experiences, are more empathetic then V/Cs and banks, and will provide more pertinent, applicable aid and assistance when needed. The management team's fear of ownership erosion, when development problems arise, can also be much alleviated.

The investing company benefits by receiving early access to the product, possible enhancement to their present product, a new product line, and future equity value in the company, product, or both.

Informal Equity Investors: The Angels

It is possible to establish and develop partnerships with local wealthy individuals or covens of the wealthy, who will put up the monies to fund companies and their product development. To accomplish this, someone in the company needs to have access to these angels. If you're a community unknown, success is probably not obtainable, unless your exemplary reputation has preceded you. It's worth the effort to try.

The first company I started was funded in this manner. In this instance, the angels found me through a mutual business acquaintance. I had networked my contacts and let the word out that I had a start-up company that was searching for investors. Lo and behold, not one, but two interested angels called. Not an accidental accomplishment; but I sure felt blessed. Three years later this company successfully went public.

Although this method of funding worked exceptionally well for me, I believe that the synergistic partner route is easier to establish and control. And, unless your angels happen to be accomplished individuals in your industry, the synergistic partner's method will provide many more tangible and intangible benefits.

Research and Development Ventures: Asset Sharing

There are a variety of R&D venture models. Two possibilities are: the R&D Partnership and the equity partnership. Either can aid companies by funding the development and sharing the risks. Although, in certain instances, tax benefits for the investor may be likely, the real advantage lies in the ownership rights to the product. It is essential that investors and companies alike review current federal and state tax laws before entering into such arrangements. A legal type proctor should also review the agreement.

Help in forming such relationships can be obtained from a variety of accounting and consulting firms. Basically, the following documents are required:

- Partnership Agreement. This document sets forth the terms for sharing profits, losses, and distribution.
- Technology License. This license specifies that the investors are granted the right to use the base technology.
- Royalty License. This license sets forth the royalty rates, limits, payout periods, and other terms.
- Joint Operating Arrangement. This agreement describes arrangements following development of the product; for example, it defines who will do the manufacturing and marketing.
- Buyout Arrangement. This agreement defines the options to purchase the technology following the development.

The R&D Partnership (RDP)

In an RDP, a limited partnership is formed with either individuals, such as company founders, or a corporation as general partner. The general partner contributes the product while the limited partners contribute capital to finance the product development. When the product is successfully developed, the general partner and limited partner share in the income generated from sales of the product. In some arrangements, the partnership will sell all rights in the product to the general partner in exchange for royalties based on sales.

From the general partner's viewpoint, the option sale shifts the research risk to the partnership, as the general partner need not buy the product rights, if the research doesn't pan out. The general partner also retains control of the technology.

The Equity Partnership

In an equity partnership (the partnership form most often used by synergistic partners), the company and the investors are co-owners of the business. The investors are generally allocated the majority of the losses, including R&D expenses (check current applicability) and the company manages the partnership.

For the company, the advantages are less dilution, flexibility in obtaining subsequent financing and control of the operations.

Profits are shared by the two entities in proportion to their ownership percentages. Development of early-on arrangements, which ensure that adequate funds are retained by the company to continue its business and develop additional products or product enhancements, is imperative.

What to Avoid

Things to avoid in any partnership include:

- Unlimited reporting and meetings which erode the company's efficiency advantage.
- At all costs, creating an adversarial relationship.
- Allowing the investors to run the operation. The general partner is the company that expends the efforts to develop the product and not the entity that only spends.
- Creating or allowing the emergence of an oppressive, domineering, and stifling partner. Both companies need to keep in mind the objective of the partnership at all times.

III

Planning: The Groundwork

"Would you tell me please, which way I ought to walk from here?"

"That depends a good deal on where you want to get to," said the Cat.

"I don't much care where." said Alice.

"Then it doesn't matter which way you walk," said the Cat.

— Lewis Carroll

Traveling down the right road requires reading from the right map. Our map is the business plan — your passport to the investment process. The following sequence describes how to design the map before building it; a logical sequence with which the cat would likely agree.

The business plan will be divided into two separate documents: the plan itself and the appendix. The plan, or portions of it, will function as the investor's introduction to your company and the investment opportunity. The appendix will contain supportive and more technically in-depth information, and should not be dispersed until an investor exhibits real company interest.

Overview of the Plan and Appendix Documentation

Reference the sample business plan in the Addendum. Note: An asterisk (*) identifies line items found in the Appendix material.

Executive Summary

- The summary, designed to stand alone, presents an overview of the company: its market, product, key management personnel, and financial condition and needs.

Management

- A summary of each member's background and qualifications.
- A description of management's complementary strengths.
- Profile type résumés.*

Marketing

- An overview of the marketplace, market size and the market niche.
- A marketplace report — preferably by an outside firm; but, at the very least, a complete summary of accumulated data from market studies done by outside firms.
- Graphs of market growth for five years.
- A description of the product's benefit to the user.

Product

- A product description overview describing the product's uniqueness.
- Block diagrams of the system with summarized theory of operation.
- System performance.
- Present status and development schedules for an evolving series of products.
- In-depth product description.*

Product and Marketing Positioning

- Product marketing schedules.
- Assessment of the competition.

Sales

- A description of how the product will be sold.
- Schedules and sales forecasts.

Operations

- Definition of how manufacturing, product support, service and quality control will be conducted.

Financial

- Assumptions: sales, cost of sales, departmental expenses in percentages, payable's time period, receivable's time period, depreciation, extraordinary events, and debt types.
- Summary: sales, cost of goods, operating expenses, net income, market size — yearly, for five years.
- Income Statement: monthly for two years, then quarterly for three years.
- Cash Flow Forecast: monthly for two years, then quarterly for three years.
- Balance Sheet: monthly for two years, then quarterly for three years.
- Personnel requirements: monthly for two years, then quarterly for three years.*
- Capital equipment requirements: monthly for two years, then quarterly for three years.*
- Cost of Goods Sold: monthly for two years, then quarterly for three years.*
- Operating Expenses: monthly for two years, then quarterly for three years.*
- Budget breakdown per department: yearly for five years.*

Investment Opportunity

- Description of investment opportunity.
- Return on investment analysis — present value method and replication method.
- Breakdown of funds needed.

Who Does What?

The following section describes the what, the who, the how and some of the why of completing the subsections of the business plan. There exist many canned, write-by-number plans which can be purchased; beware — you don't want to appear to be a company mock-up with mannequin managers and a punched-out product. The competition for investment dollars is tremendously high and it takes little intelligence to understand that uniqueness is better. Use the attached business plan, in the appendix as a reference, not a template.

Orientation

The overall goal is to obtain funding. The plan, therefore, should be directed at the funding providers — the investors. Sounds simple, but most plans aren't. Many fall prey to the sacrificial, product-love syndrome. It's imperative that you aim the plan and all the subsections at the intended reader — the investor; not thyself!

In declining order of importance, the investor will weigh the company and management team first, the market to be served and how the company will serve it second, and then the product. The product is even less important if the company does not yet have a prototype and much development work still remains.

Schedule

Assuming time is of the essence, the management team should next create a time and task schedule for compiling and completing the plan and the appendix.

Management Leader Tasks

If the marketing leader is not yet on board, then the management leader compiles and authors the plan's marketing section, in addition to the company and financial sections. If a financial leader is also on board, he/she can complete the majority of the upcoming financial sheets. Whether the management leader completes the financial section or not, he/she should thoroughly understand each financial sub-section. If not, severe embarrassment awaits the unknowing leader at the first investor's meeting chaired by the knowing.

It is also the management leader's responsibility to coordinate the plan's development and ensure conformity to the company's objectives. There's lots of hard work ahead, leader; but, as Nietzsche put it, "What doesn't kill you, makes you strong."

Executive Summary Section

The Executive Summary will be created after all of the following tasks are completed. For now, summarize the company's overall objectives so that the team gets a clear indication, early on, of where the company is, where it's going, and how it plans to get there.

Management Section

Create key personnel profiles. Make sure that you use the team approach. Team members which complement each other present a congenial, united posture to investors. Set superlatives aside and emphasize how each members' individual specialties will complement the others to the advantage of the company.

Financial Section

Projected Income, Cash Flow and Balance Sheet Statements are included in the Addendum as references. These financial sheets assume that sales start in the tenth month, which follows the completion of a customer acceptable product, preparation for production, development of marketing and sales departments and literature, and development of the company structure and moving into new facilities. Note: The financial examples in the Addendum should be used as guidelines only. If you simply copy these financial sheets you may end up with a synthetic portrayal of a company, which you little understand and can little explain.

There are a variety of financial support charts and worksheets that need to be completed before the formal financial statements are tackled. Proceeding step-by-step through the following worksheets will help you assimilate and accumulate financial knowledge at a gradual, progressive pace. Upon completing this section, you'll know what each financial line item means and contains. Compilation and insertion of these documents into the financial sheets will follow.

SALES: Emerging and start-up companies beware! Restrain your enthusiasm and create realistic financial sheets. If your marketplace is expected to be $10 billion and you assume the company can capture two percent of the market by year five, you're predicting that the company will have a conservative value of $200 million in year five, assuming 10 percent profits and a price earnings ratio (P/E) of 10. (Don't let these numbers and terms be intimidating. You too will be able to calculate these values shortly. If you want a preview, refer to the Return on Investment (ROI) Analysis section within this chapter.) It is immaterial whether or not you believe that your new company can achieve a valuation of $200 million in five years, the problem is most investors will not and unless you're asking for at least $10 million in investment funds, adequate investor's equity interest in the company will apparently not be there. Why? Knowledgeable investors will: rely on their past experiences with other new or emerging companies and not on what you assume to be possible; investigate how other companies in your marketplace have performed over similar periods; look at your past performance and your level of experience, and; be swayed by present economic trends. There's very little here that you can influence. In addition, if your company is a start-up or emerging company, the apparent risk of failure is quite high. Because of this high risk, investors will normally require a ten times return on their investment in five years. At a company valuation of $200 million in five years, the present value of your company would be approximately $20.3 million. If you were asking for a $3 million investment, the apparent investor ownership would be equal to $3 divided by $20.3 or 14.8 percent. This percentage of investor ownership in your company, for the amount of risk, would not be attractive to any investor.

Try $60 million in sales in year five with after tax profits of 11 percent plus, at a P/E of 12 for an investment of $3 million (this example is used throughout the remaining sections of this book and in the sample business plan). Prorate your financing needs using equivalent ratios in relation to your company's estimated sales. Although the example's investor ownership ratio is still too low, allowances have been included to provide some working area for the nonbelievers to reduce your sales without increasing their ownership beyond reasonable levels.

Create a Sales Forecast Chart with sales on the vertical axis and months and years on the horizontal axis. Start with the product development schedules or the beta-site schedule if the product is complete, and determine when sales can be launched. Draw a sales curve from the launch date ($0 sales) to your estimated year five sales. Developing sales forces or distributor networks will take time and these inefficiencies will change the shape of your curve. Passing through certain sales boundaries will also take longer than expected, due to the likelihood that you will be unable to add the right personnel at the right time. Although, a straight line may be the shortest distance between two points, because of these uncertainties, your line better look more like a laid over, smooth S. (Turn to the Addendum's business plan sales section to view an actual Sales Forecast Chart.) Finally, refine the curve with more accurately projected first year sales.

Don't be concerned about the rough approximation used here; you will be refining this curve throughout the business plan's creation and development and forever more.

Cost of Goods: Develop a Cost of Goods Sold chart monthly for the first two years and quarterly thereafter. Include manufacturing department costs, parts and outside manufacturing services. If manufacturing is to be done outside, obtain quotes from these type service companies. Determine if they will also carry the inventory; many will.

Personnel: With the technical and marketing members of the team, determine personnel needs per job classification and department for the first two years. For years three through five, pick off the annual sales from the sales curve and project personnel requirements as follows:

- If manufacturing and service will be done inside, divide the sales by 100,000.
- If manufacturing and service will be done outside, divide the sales by 150,000.

The divisors are approximate sales per person in an efficiently run company. (More accurate personnel numbers can be obtained for years three through five, by calculating your industry's average efficiency rating. You can determine this rating by obtaining financial reports from companies within your industry and dividing their yearly sales by their total personnel. You can also obtain these ratios from a variety of consolidated public company reports located in your local library.)

The result is the total personnel required at the end of year three, four, and five. Actual personnel requirements, of course, do not take radical jumps at year boundaries and you will have to extrapolate the proper quarterly numbers. Now would be the ideal time to start using spreadsheets with formula capabilities.

Lay out a yearly salary compensation chart per job classification and department for five years. Provide annual percentage increases for those employees on board over a year. This chart combined with the personnel requirements chart will produce the operating expenses' salary line item.

I did all the financial sheets by hand for the first company I started and, although I gained an unbelievable amount of knowledge, there was much wasted time doing mundane things like equipment depreciation for five years. With a strong recommendation, you should invest in a computer and good spreadsheet software.

Now is also a good time to take a break and understand why you're doing all this work. Historically, there are two primary ways (excluding inheritance and winning the lottery) for you to make a fortune; in real estate or by having ownership stock in a company. You've selected the second route and if you properly prepare and do this smart, you too could be among the fortunate.

FINANCIAL ASSUMPTIONS: Develop the plan's Financial Assumptions document following the sample guidelines in the Addendum. Since all of the financial sheets, after year two, will simply be based on percentages of the sales forecast, it's a good idea to also develop an in-house-only assumptions document. This informal document would, for example, contain your formula for determining personnel requirements during the years three through five. It can also contain generalized formulae for determining: how much and what to add in expenses and equipment for each new employee coming on-board; what percentages to apply to the cost of goods; operating expenses broken down by department; and pre-tax income after year two. Of course, these percentages have to "flow" from year to year — the financial projections should not exhibit "hiccups" as they pass over year boundaries.

It's the management leader's task to create and rework the formal and informal assumptions document until the percentages make sense. For example, if the entity is a product company, its R&D expenditures should approach its industry's average R&D expenditures by year five.

OPERATING EXPENSES: Develop an Operation Expenses Chart, categorized by department, monthly for the first two years and quarterly thereafter. Items to expense per company departments are:

- *Administration and Finance Department*: Salaries, payroll taxes and insurance, health insurance, business entertainment, travel, vehicle, a percentage of the company's total rent and utilities, depreciation on department's equipment, new employee recruitment and relocation expenses, legal and other outside services, maintenance, supplies and minor equipment. Note: Capital equipment costs are not expensed items.
- *Marketing Department*: Salaries, payroll taxes and insurance, health insurance, business entertainment, travel, vehicle, a percentage of the company's total rent and utilities, depreciation on department's equipment, new employee recruitment and relocation expenses, company and

product advertising, conventions, documentation, brochures and printing, other outside services, maintenance, supplies and minor equipment.

- *Technical Department*: Salaries, payroll taxes and insurance, health insurance, travel, vehicle, a percentage of the company's total rent and utilities, depreciation on department's equipment, new employee recruitment and relocation expenses, other outside services, maintenance, supplies, and minor equipment.
- *Manufacturing Department*: Salaries, payroll taxes and insurance, health insurance, travel, a percentage of the company's total rent and utilities, depreciation on department's equipment, new employee recruitment and relocation expenses, other outside services, maintenance, supplies and minor equipment. Expense pre-manufacturing costs, like prototypes, to the technical department and actual manufacturing costs to Cost of Goods.

You can spread the operating costs, like rent and depreciation, to each department, or determine overhead costs and apply to each employee — either method works. Departmentalizing the costs provides a better indication of where monies are being spent and where costs need to be cut. Remember to periodically increase rent and associated costs as you outgrow the facilities.

The following Statement of Projected Income, Statement of Cash Flow, and Balance Sheet will be included in the plan. Other financial support documents, such as Cost of Goods, should be inserted in the appendix.

PROJECTED INCOME STATEMENT: Generate the Projected Income Statement, monthly for the first two years and quarterly thereafter. A sample Statement of Projected Income is included in the Addendum and is self explanatory, assuming you've completed the preceding.

CASH FLOW STATEMENT: Cash is dear to any company, but, for the emerging company, cash is king. The Cash Flow document, therefore, is the most important financial chart, for now and until the emerging company is profitable. It is an absolute requirement that the management leader understands each and every aspect of this document. The sample Statement of Cash Flows in the Addendum may be used as a guideline. There are a variety of cash flow layouts available; none will please all investors. A brief explanation of the line items follows:

- *Accounts Receivable*: Cash received from sales is entered at the time the cash is received. The Financial Assumptions document should define the delay from sale bookings to cash received from that sale. The Addendum's example assumes a sixty day delay.

- *Cost of Goods*: Cost of Goods is entered at the time the cost to produce the product is paid. Any abnormalities should be defined in the Assumptions.
- *Capital Equipment*: Determine capital equipment requirements per department, monthly for the first two years and then quarterly. Although equipment is not an expensed item, as it has value extending beyond one year, its cost is entered as it is paid for. The Assumptions document should also define when this event occurs.
- *Operating Expenses*: See the Financial Assumptions document in the Addendum.
- *Income Taxes*: Income taxes are paid quarterly, following the offset of loss write-offs incurred previous to becoming profitable.
- *Sale of Stock*: This line reflects the funds received from investors to offset the cash disbursements. In the example, $800,000 is utilized up to the initialization of sales, and $2,097,000 thereafter. The Balance Sheet reflects the ownership costs for receiving these funds. No investor loans are assumed.

BALANCE SHEET: Now for the #&%*! Balance Sheet. You may need help here! A line-by-line explanation follows and a sample Balance Sheet is included in the Addendum for reference. Recognizing that the Balance Sheet is not a cash flow document will help.

- *Assets*: Cash is good and is thus an asset. Enter the Statement of Cash Flow's ending cash balance here.
- *Accounts Receivable*: This number is derived by accumulating the monthly net sales (see Projected Income Statement) and subtracting the cash received from these sales (see Cash Flow Statement). The method of recognizing receivables follows the guidelines set forth in the Financial Assumptions document.
- *Prepaid Expenses*: Items such as workers' compensation and liability insurance go here. This line item is usually not included in forecast plans because the amount is comparatively minuscule.
- *Property (Equipment)*: Property is recognized as a liability and is entered cumulatively as it is incurred. The financial assumptions define the payment terms and booking method for property. Use your

Capital Equipment Worksheet to determine property amounts. Reduce the equipment costs by its accumulated depreciation (the property loss in value over time). See the Cash Flow Statement's depreciation expense.

- *The Deferred Tax Asset*: (tax offset asset due to start-up losses): This is entered accumulatively and is derived from the Income Statement. This asset declines in time to zero as profits from sales revenues (taxable income) are realized. It may also be affected by partnership arrangements, as tax write-offs are generally allocated to the investing partner during the development phase.
- *Liabilities*: Accounts payable follow the guidelines set out in the Financial Assumptions document. Payables, which are paid within the month incurred, are not entered here. Remember that the Balance Sheet is not a cash flow document. Income taxes are entered from the Statement of Projected Income at the time that they become a payable item — after the deferred tax asset has been completely recognized and in the quarter following each profitable quarter.
- *Stockholder's Equity*: Capital stock reflects the accumulative capital paid into the company by investors. The company, in turn, provides investors with company ownership stock.

To obtain retained earnings subtract the sum of paid in capital stock and total current liabilities from the total assets.

Your Balance Sheet balances if the total liabilities and stockholder's equity equals the total assets.

FINANCIAL PLAN SUMMARY: Generate a one-page yearly summary, listing in order: sales, cost of goods, operating expenses, and net income. List the yearly market size, market share, and product introductions beneath the financial summary to tie the financial performance with the marketplace opportunity.

INVESTMENT OPPORTUNITY: Describe the opportunity and the projected return on investment (ROI) that result from the company obtaining its goals. For investors, the ROI is the primary reason why they would fund your company. Thus, the return needs to meet their expectations. For the company, the least amount of ownership lost is important and this page, properly executed, will support reasonable sharing of the company. Lest we forget, remember the objective — get the funding!

The Investment Opportunity write-up is essentially an indication that you've done your homework. Don't expect investors to be "turned-on" by financial sheets that suggest they will receive 15 times return in five years (72 percent annualized return). What you're trying to accomplish is not to "turn-off" the investors. You can avoid this by creating a business plan which is accurate and reflects knowledgeability and a willingness to share the company equitably. High investor interest will most likely occur if they get excited about the product's marketability and they like the "team."

How much of the company will you have to share? The percentage of equity that investors require is directly proportional to the apparent risk. Typically, inexperienced management, new and unproven products or marketplaces, and long product or market development times are examples of apparent high-risk investments. Investors, in these cases, will request a greater share of the equity to compensate for the greater risk. The following Return on Investment Analysis can provide you with an estimate of the percentage of company ownership that an investor will obtain in exchange for their investment in your company.

RETURN ON INVESTMENT (ROI) ANALYSIS: The following two methods of calculating ownership percentages both assume a company with sales of $60M in year five, after tax profits of 11 percent of sales in year five and a price earnings ratio of 12. You should use the P/E ratios consistent with established companies in your industry. A friendly, helpful broker can provide these ratios.

FIGURE 3.1: Present Value ROI Analysis

	$M
Projected sales in year five	$ 60
Investment needed	3.0
5th year profits = 11% of 5th year sales = 60 x 0.11 =	6.6
Company valuation in year 5 = (P/E) x (5th year profits) = 12 x 6.6 =	79.2
Present company value = (company valuation) ÷ (1 + i)n = 79.2 ÷ 9.85 =	8.0
i = 58% (10 times return), n = 5 (in 5 years)	
Investor's company share = (investment) ÷ (present value)	
= 3.0 ÷ 8.0 =	37.5%

- *Present Value ROI Analysis*: is a method of determining ownership percentages based on the projected future value of the company. The ratio of the present company value to the funds invested provides an approximation of the investor's ownership in the company. Figure 3.1 (Present Value ROI Analysis) assumes a 10 times investor return (58 percent compounded yearly) in five years on an investment of $3,000,000.

Unless asked for, the investor's company share percentage should not be shown to the investors. They are experts at determining these ratios — don't insult them by doing their homework!

REPLICATION ROI ANALYSIS: This is a method of determining ownership percentages based on the time expended for a new company to recreate your existing work product. When the "me too" company reaches your present position, your company will have a value displaced forward by that period of time. Figure 3.2 (Replication ROI Analysis) assumes a 10 times investor return (58 percent compounded yearly) in five years on an investment of $3,000,000.

If your product has been developed the replication method will best indicate to investors your company's real value, will help alleviate some of the

FIGURE 3.2: Replication ROI Analysis

	Months
Concept stage	12
Study and product definition phase	4
Product and company development	28
Total Months	54
	$M
Annualized profits of your company at month 54	5.7
Company valuation at month 54 = (P/E) x (profits) = (12) x ($5.7M) =	68.4
Present company value = (company valuation) ÷ $(1 + i)^n$ = 68.4 ÷ 7.83 =	8.7
i = 58% (10 times return), n = 4.5 (in 4.5 years)	
Investor's company share = (Investment) ÷ (present value)	
= 3.0 ÷ 8.7 =	34.5%

investor's early competition concerns, and will provide a supportive reason for the company to retain more ownership. If the company has not yet developed a product, the present value method will have to be used, with the fifth year sales greatly reduced due to the delay in starting sales. With exceptions, such as previous successes, existing sales, don't expect the company to be able to retain over 50 percent ownership — much less is the norm.

Breakdown of Funds Needed: Create a generic document which identifies how the new funds will be utilized. List cost items, such as fixed overhead, marketing and sales development, manufacturing and service, and next product development expenditures. This breakdown will not be part of the plan or appendix; but, you will need the information at hand to answer probing questions.

Technology Leader Tasks

Knowing the intricacies of the product is not enough. The team's technology leader must also exhibit advanced knowledge in project development and technical personnel management. He/she needs be able to clearly explain and present the product to laymen as well as fellow technologists. In addition, the technology leader needs to understand and interweave the overall company objective throughout the product descriptions.

Generally, the first plan readers are not technology whizzes and if you don't get by them you don't get. In other words, if they can't understand it, they won't invest in it. Have the marketing leader proof the copy — if he/she can understand it, it's probably OK and he/she may be able to add a little pizazz. Getting investors excited about the product is easier to do if the product's portrayal is tightly tied to market and customer needs.

- Describe the product's uniqueness and proprietaries in terms that would be understood by investors. Do not use mnemonics or abbreviations such as, OS/2 without first stating, for example, "the software Operating System (OS/2) for IBM Personal Computer (PC) compatible computers."

  As an electronic engineer, I had difficulties determining what was too technical and I have great empathy for the technical trying to explain to the temporal. The technology leader needs to perform a balancing act between too deep (no one can do it) and too easy (anyone could do it). It's a good idea to create a little mysticism in the write up; but, again, if they can't fathom it, you sink.

- Create block diagrams of the system with a summarized theory of operation. It is not necessary — nor smart — to reveal any secrets or proprietary properties in detail.
- Describe the system performance. If your product is better because of its efficiency and economics, emphasize those characteristics.

  For emerging companies, their products need to be better than what exists; if not, they don't have a product that would interest early-on investors. Producing "me too" products may make some sense for established conglomerates, but not for emerging companies.
- Detail the present status and development schedules for an evolving series of products. To the investor, a one-shot product company would likely expire before they could liquidate and cash-out. The product evolution should also not be discontinuous. A company planning on producing a multiplicity of product types has little hope for success and survivability — keep yours focused on a market niche.

  The previous write-ups will be included in the guts of the plan. The following descriptive, technical dissertation will be inserted in the separate appendix.
- Describe, in glorifying detail, the wonders of the product. The reader, in this case, is assumed to be a highly competent technical guru. Again, it's best not to reveal in detail anything that's proprietary. Later, when you're nearly there, in a one-on-one presentation meeting, the product's proprietary aspects can be discussed.

Marketing Leader Tasks

Knowing how to sell is not enough. The team's marketing leader needs to also exhibit advanced knowledge in market development, product positioning and personnel management. The entrepreneurial marketing leader is different than most other marketing personnel. He/she has control of the ego and can sacrifice present salary requirements for future larger rewards. This major team player will convert the greatest widget in the world (see Technical Leader section) to the state-of-the-art market write-up. Most importantly, the experienced marketing leader will ensure that when the plan is complete, it is salable and professional in appearance.

No fancy show biz job, please!

Marketing Section

- Create an overview of the marketplace, market size, and market niche. Work in the product's compatibility and conformity to the marketplace niche.
- Describe the benefits to the user of the company's products. Investors want evidence that the management team has a clear idea of who will purchase their product and why.
- Create lots of charts, including market size for five years and projected sales for five years, and indicate on the bottom of each the source of the information.
- Use outside services, if present funds allow, to obtain pertinent marketplace studies and competitor status. If funds are a problem, you will have to dig out this information from trade journals, competitor advertisements, contacts, and by "sweet talking" market study groups. It has to be done. Investors want to know if there are any mice out there for your better mousetrap.
- Describe how the product will be sold. If direct sales are planned make sure that the financial section is adjusted to reflect the lengthy time period required to develop sales offices and staff. If indirect (reseller) sales are planned the sales revenue should reflect the product's discounted price, not the list price.
- Using the technical leader's product development charts, determine launch dates for the product and create marketing schedules and sale forecasts.

Competitor Section

If the plan is to be taken seriously, inclusion of a realistic appraisal of your company's competition is required. Failure to be forthright will make it appear that you're either ignorant of the industry or not confident in the product's market capabilities.

- Obtain information from sales literature, customers, purchasing agents, sales reps, trade journals, and marketing consulting firms. Describe the competition and explain why your company and product are better.

Note: This section may be merged with the marketing section, if it flows better for you.

IV

The Business Plan Format

The business plan, the guide-book with futures, creates the first impression of your company and its leaders. Its appearance, grammar and organization mirrors the principals' apparent character, professionalism, and discipline. Do it right — make it look like it represents dedicated individuals who will diligently treat the investment with care!

Don't use offset printing and expensive binding. A PostScript laser printout with a simple clear plastic cover is adequate. Substance will count; lavishness will discount.

Putting It All Together

Collect and assemble the pieces of the plan and arrange as outlined below.

1. Cover and Title Page
   Title, copy number, proprietary non-copy statement (see the Confidentiality section), company name, address, telephone number, fax number and Internet address.

2. Summary
   A three-page summary describing the company and explaining who the principals are. This section should include a summary of the company's present status, the marketplace, the product, the reason for developing

this product, and the investment opportunity and capital sought with brief financial results. Insert at least one quote from an industry expert which obviously supports your product and marketing positioning.

The summary is the most important section of the plan. It should contain enough information to stand alone. In many cases, a letter with attached summary will be the only document the investor receives. Even if investors receive the entire plan, if they don't like what they see in the summary, out it goes. In general, the summary has five minutes to excite and interest the reader. For V/Cs, strive to make the first sentence sell the deal. Emphasize the marketability of the company's product and the dynamic growth market. Add excitement!

3. Table of Contents
   Complete index with page numbers.

   Note: It's preferable to send the Summary with the Table of Contents to unknown investors; not the whole plan. They get the salient points of the deal in the Summary with proof that there is more to come if they're interested.

4. Management
   Profile type résumés; one page per team member.

   Résumés without dates, but lots of substance are best. The investors want to see evidence that the management team has the experience and history to make the venture successful.

5. Marketplace
   Identify the marketplace, niche, size and opportunity, backed up with hard evidence. Use your previously developed charts.

   This section will make or break the plan. Convince the investors that your company has targeted an attractive market and has developed a plan to capture an unfair share of it.

6. Product
   Describe the market-driven product, how it matches customer needs, its present status and basic operation, cost, performance, and future evolution with development schedules. Make it understandable for the intended reader.

7. Product and Market Positioning
   Describe and analyze the competition and how your product is positioned to win.

8. Sales Strategy
   Define the selling methodology and how it matches your company's philosophy and resources.

9. Operational Plan
   Describe the manufacturing concept, the service policy, and how the product will be supported.

10. Financial Plan
    Insert your one-page summary first with the Assumptions following.

11. Financial Reports
    Projected Income Statement first, then the Cash Flow Statement followed by the Balance Sheet.

12. Glossary
    If the product is highly technical, provide a courtesy glossary to define any technical terms used in the plan.

The Appendices Format

The Appendix should be under a separate cover with a title sheet similar to the first one. This supporting assemblage of documentation may be used when the company reaches the due diligence stage with investors.

Collect and assemble the pieces of the Appendix and arrange as outlined below. Any other documentation that supports the plan and would aid in the due diligence task can also be inserted here.

1. Engineering and Product
   Insert the in-depth technical document here.

2. Marketing and Sales
   Insert any marketing studies or forecasts supporting the plan.

3. Financial Reports

   - Operating Expenses
   - Capital Equipment
   - Cost of Goods
   - Departmental Budgets, derived from Operating Expenses and Capital Equipment.

V

Finding Investors: The Quest

Sources

There are endless sources of funding available, but not all of them are worth the effort. Broadcasting your business plans to hosts of interacting venture capitalists probably wouldn't be a very good idea. The investor types listed in this section, therefore, are those that make sense: quality sources with a higher probability of success.

This section also assumes that your funding requirements exceed $500,000. If you're searching for a relatively small sum of start-up dollars, say under $250,000, other sources are available, such as state funds, the Small Business Innovation Research Program (SBIR), Small Business Investment Companies (SBICs), community development funds, and the National Business Incubation Association.

Investors are usually divided into groups based on preferred dollar range of investment, preferred type of business or industry, and location. The largest number of plan rejections occur because entrepreneurs send plans to sources who do not fund their type of company. Save time and do it smart — research and then select compatible investment sources.

Send the plan to a few qualified, compatible V/Cs and then spend most of your efforts contacting synergistic candidates. Call the contact before mailing a plan. A proposal sent in the mail without a phone call is DOA.

Referral calls can make a difference — an introduction from a recognized associate works the best.

Since the initial phone call is your introduction to the investor and vice versa, and you have about 30 seconds to whet his appetite, it is essential that you prepare. Future auditions will depend on how well you can sell yourself, the team, the company, the product marketability, and the future in half a minute. One of the keys is to let honest enthusiasm flow with vigor and excite the investor to act. No taffeta phrases, just ardency with punch. Get to the point and to the next step — scheduling the mailing of the Summary or plan or setting up a meeting — quickly.

This is one of those practice-makes-perfect events — don't be discouraged by the initial results.

Stick-to-itiveness is now a required management asset — it will take time to find compatible investors. Emerging companies be forewarned that V/Cs will rarely tell you no: don't let them drag out the decision-making time until you're mortally wounded. One of their favorite final replies is, "Our investors usually do not fund early companies like yours, but I'm sure they would like to be involved in the next funding phase." This means no and we'll call you.

Venture Capitalists

Your probability of success depends on the company's status. For emerging companies, the probability of obtaining funding from V/Cs is low. For more established companies, V/C funding is probable.

In general, V/Cs today aren't looking for seedlings, they're searching for mature oak trees, sans the root fungus.

I would suggest you obtain a guide to venture capital sources from your local library or book store, for a listing of V/Cs, the locations which they cover, and their preferred industries.

Private Placements

The probability of success is medium. Private placements usually come from one of three categories: wealthy individuals, businesspeople, or institutions, such as insurance companies. In most instances they are interested in low-risk ventures which they can easily understand. Sources are available via your own contacts and by networking, possibly from your state commerce department and brokers. If you're not well established in the community the success probability is very low.

R&D Partnerships

Limited Partnerships: The probability of success depends on present tax laws and the amount of write-off provided. Unless the present tax laws change, this method has little chance for success. The limited partnership type of R&D partnership should not be confused with synergistic R&D partnerships. In the limited partnership, the tax shelter aspects are important. In a synergistic partnership, the product produced and the resulting sales and revenue stream override the tax benefits.

Synergistic Partnerships, Mergers and Acquisitions: The probability of success is comparatively high.

Mergers and acquisitions represent a continuing source of financing. The future parent company providing the funding also has valuable resources to aid the company, such as: facilities; accounting, legal, manufacturing capabilities, service and marketing personnel and contacts; as well as intangible assets such as vendor, bank, community, state, country, international name recognition, and in-depth business knowledge. It's a win-win situation for both companies.

Large companies that will fund other company's product developments exist everywhere. You will have to do some telephone "leg-work" and networking to locate appropriate ones, such as local companies searching for ways to diversify; potential customers or suppliers (both are excellent sources); and cash-rich, product-poor companies. Use your library's industrial classification guide to identify the companies in your industry and then contact their corporate development managers.

VI

Presentation: The Technique of Selling Carrot Seeds

Strategy

- Anticipate! Look ahead, expect, foresee and prepare. Following the meeting, the management team's consensus should be "great meeting," not just "good presentation." Surmise what questions may be asked; address the majority in the presentation, and be prepared to answer the remaining during the meeting. Don't "wing it."
- Create the presentation for the investors present. Continually evaluate the presentation for its clarity and marketing oomph. Excite the investor to act!
- Do not over emphasize or complicate the technology. For technical folks, have ready a set of overheads which can be inserted when needed. Technical charts should, at all times, support the marketing and product positioning and never need interpretation.

  Make up two complete sets of overheads; one for private placement type groups with limited technical knowledge and the other for the more technically astute investor. The presentation is always directed at the investor's knowledge base. For example, if the investor's background is in accounting, use examples which emphasize accounting applications for the product.

- Use quotes from industry experts or trade journals which obviously support your product and marketing positioning.
- If necessary, have a pre-agreed-to nondisclosure agreement signed before the presentation.

  Note: In general, the information revealed in the first presentation meeting should not contain proprietary information. Do not insist on a nondisclosure agreement if it's not essential. You don't want to create unnecessary conflicts for investors involved in other projects.
- State, at the beginning, that the presentation is designed to be interactive and to be interrupted at any time with comments or questions. If no one asks a question within the first five minutes, a team member should interrupt and add a comment or question, to purposely break up the presentation. The intent is to have the investors join in the fun and get caught up in the excitement.
- The team member best gifted with verbal communication skills should do the actual presentation. The presentation may work better if the management leader is not the presenter. The management leader can then observe the audience's reactions; interrupt when he/she sees a perplexed or blank look, or when audience participation wanes; reemphasize points that caused interest and attentiveness; and determine who the nonbelievers are that need conversion.
- Again: anticipate! Be prepared to answer tough and perhaps embarrassing questions.
- Do not hand out a copy of the overheads before or during the presentation.
- Never argue about anything with other team members. If the presenter fails to handle a question properly, interrupt with, "In addition . . . ."
- Keep thy egos on a leash. This is a team roping event not a one-man revue.
- What convinces is enthusiastic conviction!
- Before the meeting adjourns, make sure that the next step is defined and scheduled!
- After the investors have departed, have a debriefing meeting to discuss ways of improving the presentation.

Format

I hesitate to include the following brief as it is — as it may not flow right for the team members. Take all the artistic license you need to create your show, while retaining the above objectives.

- Immediately capture the investor's interest by opening with a show stopper.
- Next proceed with information about the principals and the company: Who you are; why you are; what you are, and where you're going — with a lot of conviction and enthusiasm.
- Near the end of the presentation, insert the one-page, five year Financial Summary sheet, with market size included, immediately followed by an overhead of PE ratios of companies in your industry. The PE ratios add validity to the ROI calculations displayed on the final overhead — the Investment Opportunity.

VII

Confidentiality and Nondisclosures

The following Nondisclosure Agreement is included for guidance only, and no warranty or representation that the information in fact will be protected by said document is implied.

Legal representatives should review this document to ensure the latest precedents are included.

In the example, XYZ is the company wishing to protect its information and *Company* is the group wishing to receive the information. Conclude with a signature page which both parties sign and date.

Nondisclosure Agreement

WHEREAS, it is the desire of Company, having offices at ________________________ to receive certain know-how and proprietary information, hereinafter referred to as "Confidential Information," possessed by XYZ, whose address is ________________________________;

WHEREAS, it is understood by *Company* that such information is regarded by XYZ as being confidential or proprietary to it, and that it desires to protect the same from dissemination to and use by persons not authorized by it;

NOW, THEREFORE, the parties agree as follows:

Confidential Information shall mean any information and data relating to the above know-how and trade secrets, which shall include but not be limited to technical know-how and all record-bearing media disclosing such information, which is disclosed or transmitted to Company.

In the absence of prior written consent from XYZ, Company shall neither disclose such Confidential Information to any person except authorized representatives of Company nor itself copy such Confidential Information.

Notwithstanding the foregoing, Company shall not have any obligation with respect to any Confidential Information which:

(a) was at the time of receipt otherwise known to Company as evidenced by suitable documentation;

(b) at any time becomes a matter of public knowledge or literature without the fault of Company;

(c) is made available by XYZ to anyone without restriction on the use thereof;

(d) is disclosed in accordance with express written approval of XYZ.

All Confidential Information in the form of record-bearing media transmitted to Company and any copies thereof made by Company shall be returned to XYZ promptly upon written request of XYZ.

All obligations of Company to keep the XYZ information confidential shall terminate five (5) years after the date of this Agreement.

(Insert Company and XYZ signatures here)

Title Page Nondisclosure Statement

The following title page nondisclosure statement is included for guidance only, and no warranty or representation that the information in fact will be protected by said statement is implied.

Legal representatives should review this document to ensure the latest precedents are included.

"The XYZ business plan and attachments are confidential and contain proprietary information including trade secrets of XYZ. Neither the plan, nor the information contained in the plan, may be used, reproduced or disclosed to any person under any circumstances without express, written permission of the XYZ Corporation."

VIII

Agreement: The Share-Cropping Terms

Important Reminders

- Remember the objective.
- Don't fight a battle if you don't gain anything by winning.
- Pace. Birds of a feather, flock together because they like each other's company.
- To make your fortune let people see that it is in their interest to promote yours.

Eventually, discussions stop and negotiations begin! When you reach the negotiations stage, your company joins the ranks of the fortunate. You've worked hard and smart to gain this plateau, don't now allow greed to smother this fairly rare opportunity. Approach the negotiation meeting with the attitude that you need to win this one as there may not be another opportunity. Fight hard, but fair, to retain the company objectives. Concentrate on the company's goals; your personal objectives should be so closely linked to the company's, that a company win means you win.

Negotiations are not a cut-and-dried affair — that's why they're called negotiations. Use a mixture of marketing and sales techniques to gain advantages; predetermined, defined strategy, with seat-of-the-pants maneuvering.

Separate the legal issues from the business issues. The initial negotiations should cover business issues only; so much so, that the presence of a lawyer should not be necessary.

Lawyer involvement in business decisions should be kept at a minimum. They should be retained for what they've been trained for — keeping you, your associates, the deal, the proposal and the company "clean." Have your attorney review the agreements before you agree and sign; not before you hash-out the basics.

Try to keep the deal as simple as possible and don't change the basic deal once you've reached agreement. Be flexible, but don't give in too easily. When you do yield, try to gain something in return.

Note: In negotiations, you are basically jockeying for position. Always insert sections in a contract that you know you'll have to eventually delete or modify. Surrender the inconsequential to capture the indispensable. Meanwhile, the fellows on the other side of the table are doing the same thing and everybody thinks they're winning. A photo finish is ideal.

Proposal

Although it is the investor's prerogative to have his or her lawyer prepare the legal documents, you should create an outline of the proposal before meeting with the investors. The following is a generalized list of those items usually covered in the contract. You should have your outline cover the non-legalistic line items, identified below with asterisks (*).

- *Description of the financing*: Authorization of securities, sale and purchase of securities, other participating investors or partners, use of proceeds restriction.
- *Company representations, warranties and disclosures*: Organization and corporate power, subsidiaries, business description, company authorization, capitalization,* financial statements warranty, absence of previous undisclosed liabilities,* absence of material adverse changes since the last balance sheet,* business plan or offering circular warranty, title to properties, applicable statutes compliancy, taxes filed, contracts and commitments existing, non-infringement of patents,* company has the right to effectuate the transactions, existing litigation disclaimer,* other agreements between principals,* insider agreements,* corporate charter and by-laws,* brokerage fee disclosure,* absence of any false statements.

- *Representations of the purchaser.*
- *Conditions precedent to closing*: Opinion of counsel for the company, opinion of counsel for the purchasers, representations and warranties certification, loan agreements, purchase agreements completed by all participants, compliance certificate, any other unique provisions.*
- *Covenants of the company*: Maintenance of corporate existence, payment of taxes and claims, legal compliance, property repair and maintenance, property and liability insurance, life insurance, employment contracts,* financial accounts and reports methodology, providing financial and operating statements, adverse change disclosure, board of directors,* access to premises, preemptive rights on future financing, use of proceeds,* arm's length dealings with related parties, nature of business not to change, no unapproved change of corporate by-laws, no dilution, redemption of securities, no unapproved sale or purchase of assets or mergers or consolidations, no unapproved encumbrances or indebtedness or dividend distribution, maximum employee compensation,* loans and advances, aging of payables restriction,* no default of agreements.
- *Covenants of the investor*: agreement to future rounds of financing.*
- *Resale of securities*: demand registrations limitations, "piggy-back" rights, form of registrations, registering allocation of expenses, investor's indemnification.
- *Other*: survival of representations and warranties, successors and assigns, notices, allocation of proposal expenses, entire agreement, amendments and waivers, state of governing law.*
- *Exhibits*: employment agreements, first right of refusal agreements, company evaluation documents.

Securities

More legalese. The agreements should detail the specific class of stock, dividend, voting, conversion privileges of the stock and any conditions concerning notes. The following are typical provisions found in debentures or notes.

- Note description: interest rate, principal, date, conversion rate
- Prepayment conditions
- Collateral

- Subordination
- Company covenants
- Defaults and remedies
- Stock purchase warrant agreement: rights, adjustments to the exercise price, notice of reorganization, definitions, expiration date, transfer of warrants, exchange of warrants.

Put considerable effort and forethought into eliminating notes of any type. Repayable loans can drag the company down at the time you should be expanding. If notes are an absolute pre-funding requirement, require that they be long-term types so that the company's current financial ratios are not adversely affected.

Apportionment

Propose a funding structure that allows equitable sharing of the company. The percentage of equity that investors require is directly proportional to the apparent risk. Typically, inexperienced management, new and unproven products or marketplaces, and long product or market development times are examples of apparent high-risk investments. Investors, in these cases, will request a greater share of the equity to compensate for the greater risk.

How much ownership? It should be apparent to all that the management team and company need to retain enough ownership to keep the team's personal objectives aligned with the company's goals. So much depends on how well the management team have sold the capabilities of management, the product, the marketplace and the company, that standardizing ownership rates is not possible. Most management leaders would be happy with a 20 percent company retained ownership after all rounds of financing are completed and the company is self-funding. The old adage "the size of the pie to be shared is far more meaningful than the size of the slice" is still right on. Of course, you don't want to end up being a pan-licker either.

Since an asking price can be lowered a lot easier than it can be raised, try to have the investors state their ownership position before you state yours. If their requested equity ownership is greater than you expected, state so and ask for their reasoning. Counter the reasons by re-presenting your case, reemphasize the management team's sweat equity and monetary investment and then state the ownership rate that you feel is equitable. You do, after all, have squatter rights — don't squander them — but also don't forget the objective.

If their requested equity ownership is less than expected, act hurt, frown, delay, look at your partners with the raised eyebrow and gracefully accept . . . with conditions. Can't ever recall this happening, but just in case . . .

Control

Operating control of the company should remain with the company managers. Try to cover requested exceptions in the Board of Director's arrangements and composition. For example, retain control of the company even though you may hold a minority position by suggesting that the board be composed of an equal number of representatives from the management and from the investors, with a neutral outside board member to serve as a tie-breaker.

This type of Board make-up has been very successful in previous companies. Try adding an outside member that could aid you in areas where you need help or guidance and or one that could assist in promoting the company.

Finance Phasing

Staged financing is the process of timing each stage of the financing to coincide with the achievement of pertinent milestones. For the investors, the risk diminishes with each achieved milestone. The investors and the management team should agree beforehand on what the consequences are if a milestone is missed.

Typical funding milestones are:

- contract signed;
- completion of the product development stage, proved through the completion of the product's first ("alpha") test;
- successful completion of customer ("beta") tests of the product;

- achieving the first sales goal; and
- reaching a growth plateau which requires expansion funding.

Staging the financing is understandably investor preferred. You can't expect the investors to shovel money in without looking where it's going. On the other hand, creating too many milestones is a sure way to create serious problems, as milestones can create adversarial relationships and sacrifice long-term goals and objectives to meet short-term performance goals. Too many milestones can also remove the flexibility required, especially by young companies, to efficiently and opportunistically react and redirect efforts to match changing marketplaces and customer requirements — the real guidelines.

In any case, it is safe to assume that the company will probably experience the usual complications and unforeseen delays all companies experience, and the milestones will need to be adjusted accordingly — within reason, of course.

Note: A true partnership is the goal, not an overseer-subordinate relationship or interaction based on ownership percentages. The two companies constructively working together as equals, with the same overall objective, can create a real winner.

Since the investor risk diminishes with each achieved milestone, the company can negotiate higher prices per share at each successive milestone completion. The company can also expect that the original investors, possibly joined by additional investors, will invest in subsequent rounds. If the company performs well, future funding should cost less. If not, expect future funding to cost more — and justifiably so!

IX

Now That You've Got It . . . Planting Straight Rows

To Spend or Not to Spend

For emerging companies cash is precious and life sustaining. Investor cash is never an unending flow. The failure of the management leader to recognize that superb cash management is an absolute job requirement has caused the collapse of multitudes of new companies. At the very least, poor cash management will always result in the management team's ownership percentage being heavily diluted. Treat the funds as you would your own retirement savings. You owe it to yourself, your partners, and your investors.

Follow the plan intelligently; be flexible — adjust the actual cash flow as events and milestones occur. This is tricky, and requires skillful balancing of future wants (reaching for tomorrow's goals) against present actual needs.

For example; if actual sales are falling behind projected sales, spending less in marketing and sales does not make sense; but, cutting back in other departments does. As sales slip, and revenue from sales is delayed, overhead continues to accrue. If you don't cut back, your funds will obviously run out before the company is self-funding. Conversely, if you cut back too much, the future growth of the company will suffer. Experience and great intuitiveness is required to carry off this delicate balancing act.

Successful entrepreneurs are born with survival instincts — it's in their genes. Additionally, the thought of failure is so obnoxious to them that they adjust, almost automatically, to changing conditions, attacking problems gleefully, while retaining their goals. If you're like that or can force yourself to be like that, you've got the beginnings of what it takes to be an outstanding cash flow manager.

Don't attempt to find solutions by throwing money or masses at problems; it's self-feeding and self-destructive. Small, tight, lean, synergistic teams will always out-produce massive, corpulent, exorbitant, pyramids of people. Historically, it's the small companies or very small divisions of large companies that have conceived, developed, and produced the majority of new technology; not the behemoths.

Large companies organized into small synergistic teams can be competitive, if they break the never ending reporting and feedback chains and eliminate job-justifying meetings.

Money management, like people management, is a skill. When you finally get the funding, treat the infusion as life supporting.

To Reveal or Not to Reveal

Schedule and budget slips and profusions of other unforeseen problems will occur throughout your company's lifetime. Reveal their existence to the investment partners and these problems can cause ownership dilution, lack of confidence in the management, and threats of intervention. Conceal and/or distort and they can become growing cancerous sores. What to do?

First, don't "choke" or panic. Problems need carefully thought-out solutions. Don't be stampeded into making an irreversible error by proposing an impracticable solution.

Second, evaluate the severity of the problem and the likely effect.

Third, determine the solution as expeditiously as possible. It is absolutely critical that this step precedes the next step. A solution must be available before you announce the problem.

Fourth, determine what levels of the organization are affected, and inform personnel at the appropriate levels of the problem, the solution, and the anticipated results. Investors need to be informed whenever schedules, budgets, and any occurrence which may affect the ROI and its time period are involved now and in the future. In some cases, you may want to, or have to, work out the final solution with the investors — be prepared to present your proposed solution immediately following the introduction of the problem.

Whenever a problem surfaces, attack. Anticipate the investor's likely reaction to the problems and always have positive solutions at hand. Never let a problem linger, fester, become infected, and ooze. Inoculate and attack!

Note: A key to long-term success is never to distort or deceive, and never present a problem without a ready, positive solution. An entrepreneur's cup is never half empty nor half full, it's always overflowing with solutions. Above all, retain your sense of humor — it's what differentiates us from other animal life and some foreign ministers.

May the luck of the work hard and smart go with you!

X

HELP

Assistance is available, if needed. Mr. Manweller's present company, Corporate Advisory, can provide experienced, professional aid to your company's management in the following areas:

- Product
  - Determining diversifying products compatible with client's capabilities
  - Discovering and developing cooperative product ventures
  - Defining new product requirements per marketplace needs
  - Assessing product compliance to the marketplace
- Marketing
  - Defining marketplace wants and needs
  - Defining steps necessary to position product and company in marketplace
  - Creating product and company promotional materials
  - Assisting in market expansion

- Assisting in public relations development

- Plans
  - Developing interrelated product and marketing plans
  - Creating/revising company business plans

- Financing
  - Conducting feasibility analyses
  - Determining compatible funding sources
  - Creating presentation materials
  - Developing meetings to connect the client to the investor

Contact

Richard Manweller

Tel.: 208-939-1101 ▪ E-mail: Coadvise@aol.com ▪ Fax: 208-939-7937

XI

Addendum

The following business plan is for data-FAST Corporation. This plan was designed to obtain financing for the introduction and production of the company's high tech database management server, DB FAST-MANAGER. Some of the technical information as well as some of the financial data has been changed in order to make the plan easier to understand. It is interesting to note that data-FAST Corporation did obtain the financing needed to introduce their new product as a result of their carefully written plan.

Business Plan

data-FAST Corporation

Copy Number ________

data-FAST Corporation • Your street address, City, State, Zip

Contents

Executive Summary 53

Management 59

Marketing. 62

Product 67

Product and Marketing Positioning 75

Sales. 83

Operations 88

Financial 91

Financial Spreadsheets. 95

Glossary 114

Executive Summary

The following Business Plan Summary is designed to stand alone. The Summary is also the leading section in the Business Plan.

In many cases, a letter with attached Summary will be the only document the Investor receives. Always include the Business Plan's table of contents with the Summary to show the Investor that you've completed your tasks and are ready to proceed.

Note: This business plan is based on an actual plan that was used to successfully obtain investment financing. Personnel and company names have been changed or dropped to protect the innocent.

Summary

"We're at the early stages of a radical change in the economics of the [computer] industry."

—*The Walker Journal*, June, 1992

The rapidly expanding, multi-billion dollar database management system marketplace is at the threshold of a new era. Dramatic advances in computing architecture, until only recently dismissed as theoretical by entrenched companies, now hold the promise of fundamentally redefining the competitive marketplace. With its powerful and innovative database server, the data-FAST Corporation will be the clear leader in this market.

Following two years of study and three years of market and product research, definition, design and development, the data-FAST Corporation is seeking financing to fund expanded production and accelerated sales. With a conservative business plan that calls for the firm to reach $60 million in sales by the fifth year (management's internal goal is $120 million), we need a strong financial partner to assist us in the funding of our growth.

Market Overview

"It's easy to figure out why organizations are downsizing their database applications. After all, corporations can lower hardware and software costs, decrease development times, reduce maintenance costs, create more intelligent applications, and obtain greater flexibility ... ."

—Robert Fernstein, Publisher, *Database Talk*, July 1993.

The computer industry is in the midst of a fundamental transformation. Driven by increasingly powerful and low-cost computing and storage technologies, general business applications, which once resided exclusively on mainframe/mini-computer platforms, are now finding broad and legitimate support in the networked "client-server" environment.

Database Management Systems (DBMS), the compute and storage intensive resource for corporate information management, is at the forefront of this accelerating revolution. Customers (particularly those in financial services, insurance, education, healthcare, transportation, and utilities) are now aggressively implementing client-server based DBMS solutions. Consequently, the DBMS market has been and is projected to continue expanding rapidly ($15.5 billion in year one to $31.0 billion by year five, a 200 percent growth rate. Source: Datacorp Market Research).

Several years into this revolution, however, DBMS customers have begun to realize that the current "general purpose" computing platforms have serious limitations, and require very costly solutions. The three key areas of concern are:

1. Scalability: as performance/capacity needs expand, customers must get a bigger box;
2. Integration: present solutions do not integrate well within their environments;
3. Price/performance: the price customers must pay for DBMS solutions remains unacceptably high and, even worse, varies by orders of magnitude across the entire spectrum of performance/capacity needs.

Product Overview

The need for a truly scalable and seamlessly integrated DBMS server in today's database management market is clear. Until now, however, no one has been able to marry the highly complex characteristics of a true relational database software model to a highly scalable hardware architecture and, in the process, meet price/performance targets. Data-Fast has accomplished this. The implementation of the data-FAST product is straightforward and uncomplicated: a highly scalable family of DBMS server products, with price/performance characteristics unmatched in the industry, aimed squarely at the client-server database management applications.

The data-FAST product, DB FAST-MANAGER™, has the following marketable advantages.

1. Minimized computer hardware/software upgrade/changeout and training expenses.

The DB FAST-MANAGER server attaches to an existing network processor, acting to off-load data management and storage functions from an in-place but "performance starved" solution. The existing and familiar DBMS remains in place, acting thereafter as the User and Network Host to the now infinitely expandable Database Server — the hardware/software DB FAST-MANAGER.

2. A safe and intelligent choice for DBMS administrators.

   Wrong guesses about present and future performance/capacity needs will no longer require hardware change-outs to the next (and more expensive) computing platform; nor will it be necessary to buy a mainframe-type system just to be safe. The DB FAST-MANAGER system allows customers to match their hardware to their present needs, and as the size and performance requirements of their applications expand, they simply add processing modules to meet demand.

3. A powerful packaged database management solution.

   Data-Fast can supply the DB FAST-MANAGER in conjunction with a relational database management system for first-time customers looking for a single-vendor environment.

4. A modular architecture and product pricing structure which allows the product to compete across the entire DBMS performance requirements spectrum.

   The product was designed with optimal modularity and price competitiveness in mind. The modularity provides customers with a single vendor solution. Data-Fast can supply system configurations from desktop units, for a stand-alone PC/Workstation user, to desk-side units, for departmental client-server environments, to towers, for the "mainframe" solution — all using the same type modules. As performance or storage requirements expand, unlimited additional modules may be added with virtually linear performance and pricing characteristics. In all configurations, data-FAST is setting new price standards in the database industry.

5. A hardware/software structure designed for optimum reliability. The product's distributed, parallel architecture is significantly more reliable than past generations of DBMS solutions. Further, planned near-term enhancements will provide mainframe levels of fault tolerant reliability.

   In addition, the product has a high competitive entry barrier. The complexity and effort which would be necessary for another company to competitively duplicate DB FAST-MANAGER is substantial and expensive. The DB FAST-MANAGER consists of copyrighted, proprietary database processing algorithms, transparent to the user.

Key Management

Management and personnel are in place to expand the company and produce and sell the product.

Richard Leader, president. Former vice president and vice president of engineering. Twenty years of experience comprised of 10 years of company management and 10 years of technical management and design. Mr. Leader has created and recovered a high-tech company, directed engineering groups in the design, development and manufacturing of high-tech products, and conceived, designed and developed leading high-tech products.

Robert Product, Ph.D., vice-president, and vice-president of engineering. Former vice-president of engineering and chief scientist. Twenty years of experience in computer systems and software and product design and development. Recognized authority with in-depth research and design experience in the database and transaction processing field.

Roger Market, director of sales. Former CEO and vice president of sales and marketing. Twenty years of computer-based sales, marketing and management experience in Fortune 500 and start-up companies. Mr. Market has established multichannel distribution systems, and pioneered the marketing and selling of high-tech software.

Other Personnel. The additional data-FAST team members have equally impressive credentials and experience levels in finance, sales, operating, manufacturing, and technical disciplines.

Investment Opportunity and Financial Performance

The database industry is the "glamour" segment of the computer-based industry. Profit margins are high and price/earnings ratios, for competitive database companies, reflect this with ratios ranging from 24 to 53. Data-FAST's financial projections assume a conservative P/E of 15.

The company is seeking $2,897,000 to fund the manufacturing and sales plan of the DB FAST-MANAGER series of products. The company anticipates that this infusion of funds will be sufficient to carry it to profitability.

Financial Summary

(In Thousands of Dollars)

	Year 1	Year 2	Year 3	Year 4	Year 5
Sales	$481	$8,995	$24,000	$44,000	$60,000
Pre-tax income	($1,032)	($1,181)	$4,248	$8,890	$13,554

Management

The Management section of the business plan expands on the summary's brief key personnel profiles. In this section, you should also include the profiles of other existing personnel that make your company look stronger.

Note: I've only outlined major management's experience. Other company personnel have been removed. You should list all personnel that are on board.

I. Management

The data-FAST management team members bring unique and tested skills to their functional areas. The highlights of their prior experience and their current functional responsibilities are presented below:

Richard Leader, president and CEO

Former president of Computer International, Inc. Under his direction the company grew to become the leading contract product development firm, with a nationally recognized name and reputation for outstanding product development expertise.

Previously, he was vice-president and vice-president of engineering at Tech Marketing, Inc., where he developed formal company forecasts and budgets, and engineering practices and procedures which enabled fixed-price product developments to be profitably completed. These changes led to the successful conversion of the company into a profitable entity.

Previously as a program manager and project engineer at Control Corporation, Mr. Leader led the team which designed, developed, and produced CC's first firmware controlled processing system — SIGHBER.

Mr. Leader is responsible for overall company operations.

Robert Product, Ph.D., Vice President of Engineering

Formerly vice president of engineering at Systems, Inc., and chief scientist at Computer International, Inc. where he was responsible for the design and development of advanced high-performance multiprocessor systems and relational database and transaction processing systems.

Previously, Dr. Product designed and developed high-level language compilers and interpreters, microprogramming for system peripherals and computers, and advanced computer system architectures.

Mr Product is responsible for the engineering department as well as design and development of new products.

Roger Market, Director of Sales

Formerly vice president sales and marketing, Products Corporation, where he was responsible for establishing a multichannel distribution system to market a microprocessor-based protocol converter/print server product line. He doubled revenues two years running.

Earlier, Mr. Market helped pioneer the selling of software as a long term member of the Formatics General management team.

He was vice president of sales and marketing for the Software Products Company. The company's sales grew from $5 million to $250 million during this time frame.

Mr. Market is responsible for the development of marketing and sales.

Other Personnel

Note: Include the profiles of other existing personnel that make your company look stronger.

Marketing

The Marketing section of the business plan contains information which supports your claim that the marketplace for the product will presently support and provide long-lasting growth for the company.

All of the company's financials and the Investor's ROI are predicted from the company's assumed capture rate of this marketplace. Therefore, your market research results should be thorough and incontestably accurate.

Note: Insert your source names beneath each chart you provide.

II. Marketing

Overview

General DBMS Environment: The computer industry is in the midst of a fundamental transformation. Driven by the ongoing development of a new generation of increasingly powerful and highly cost-effective general purpose, standards-based computing platforms, customers are now finding broad and generally effective support for virtually the entire range of their computing and information management requirements, outside of the traditional mainframe/mini computer environment. The strength and clarity of this trend is obvious when sales and forecast data for mid-range ($25K–$750K) computer systems are reviewed, as in Figure 1 on p. 64.

Key points to note in Figure 1 include:

1. The overall growth in mid-range systems — $57B in year one to $76B by year five.
2. The growing marketplace for general purpose "standards-based" systems (UNIX, OS/2 and DOS based systems).
3. The leveling-off of general purpose, proprietary system sales (DEC and IBM, predominantly).

Database Management Systems (DBMS), the compute and storage intensive resource for corporate database information management, is at the forefront of this changing marketplace. Large and small companies alike have for several years been aggressively implementing their mid-range DBMS solutions on these new general purpose platforms (UNIX, in particular) in lieu of, or off-loaded from, mainframe/mini computer platforms.

FIGURE 1: Worldwide Mid-Range Computer Marketplace — Hardware

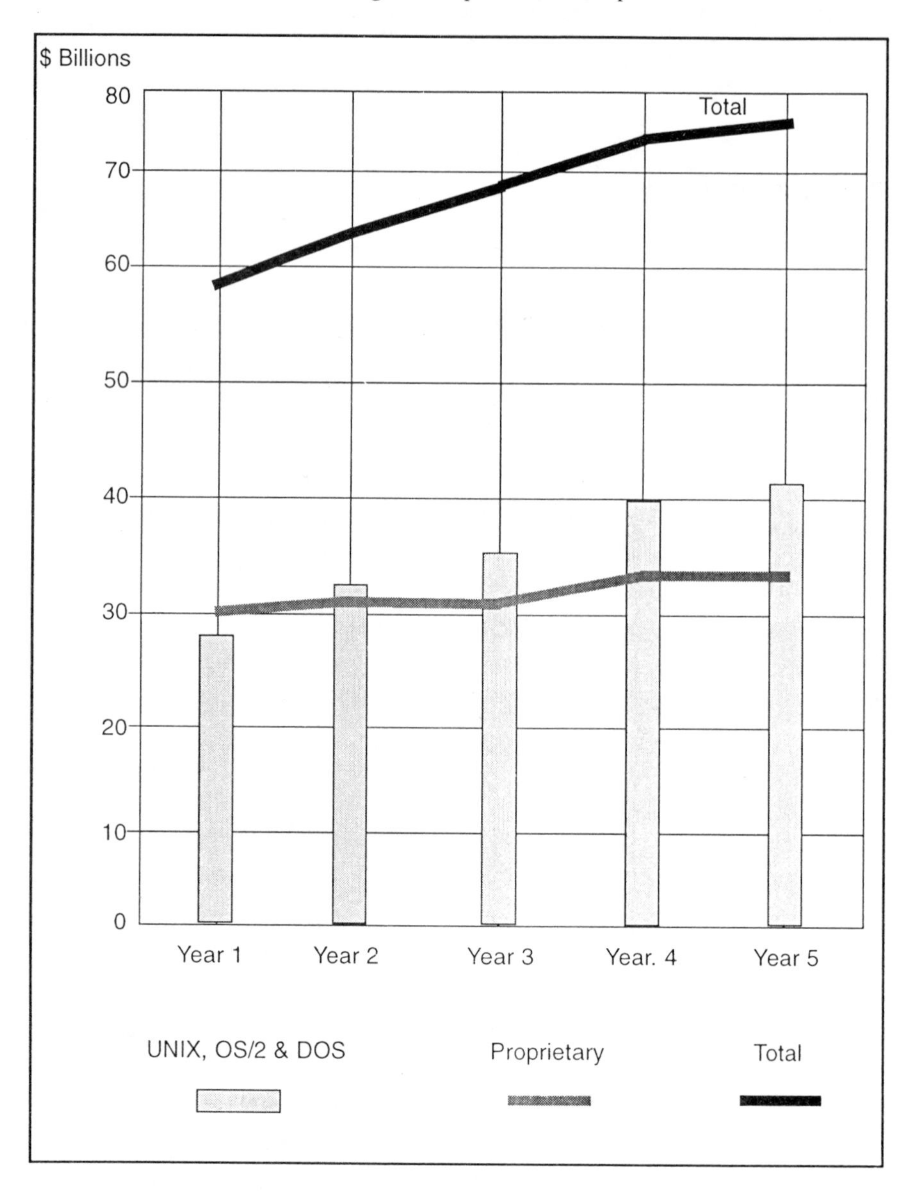

Served Market Analysis

The potential market for data-FAST's DB FAST-MANAGER is a subset of the data discussed in Figure 1. The total available market for DB FAST-MANAGER is in excess of $30 billion in year five, as shown in Figure 2.

FIGURE 2: Worldwide data-FAST Marketplace

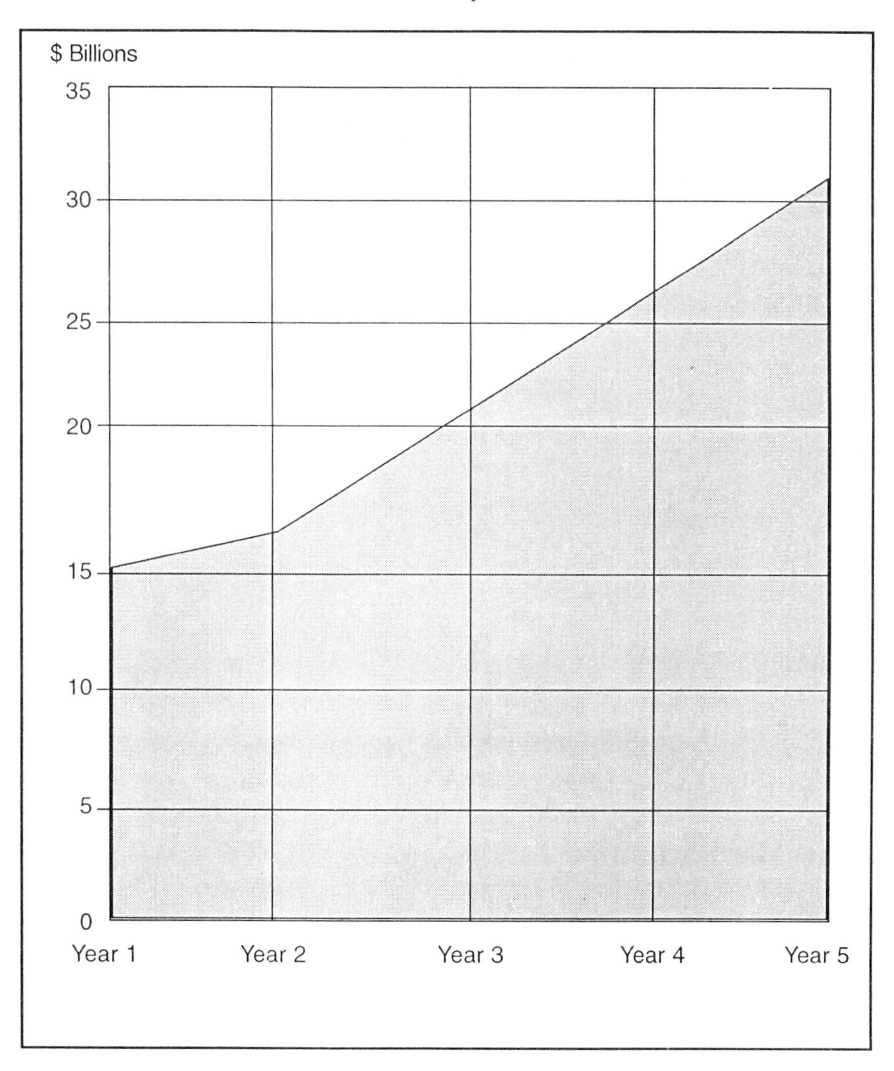

Key Customer Concerns

This initial period of evolution has been an eye-opening and two-edged experience for DBMS customers. On the one hand, a broad new range of users and organizations has been introduced to computing power and applications heretofore available only to customers with mainframe/mini class information system budgets. Users and cost-conscious executives have recently sampled these lower-priced solutions and like the investment returns. Demand, therefore, has accelerated dramatically. The other edge of the sword has begun to cut, however, as customers have begun to experience problems and voice the following concerns:

- *Scalability*: Database applications are compute and storage intensive, and their growth can be unpredictable. Customers are increasingly disenchanted by fixed-performance products which cannot gracefully and cost-effectively expand to meet the growth of their DBMS applications, and they are tired of vendor strategies of wholesale platform change-outs to achieve higher performance or capacity levels.
- *Integration*: Customers are looking for DBMS solutions which integrate seamlessly without performance impact into their existing environment. To their distress, they have discovered that their general business computing resources are often adversely impacted by the compute/storage demands of a high-performance DBMS application.
- *Reliability*: Database management is a mission-critical function. Customers need to know that their data is absolutely safe and accessible. For Fortune 1000 applications, accustomed to the security of the mainframe but beginning the transition to client-server computing, no issue is more important. Cost-effective solutions to this problem, particularly at the low to mid-range spectrum of DBMS applications, do not exist.
- *Price/Performance*: The price customers must pay for DBMS solutions remains unacceptably high and, even worse, varies by orders of magnitude across the entire spectrum of performance/capacity needs.

Product

The following Product section of the business plan contains overview information on what the product is, how it functions, its present status, and future growth plans.

Notice that the introductory subsection shows how the product satisfies the previous section's marketplace requirements.

III. Product Overview

Introduction

The need for a more effective DBMS solution in today's network computing information management model is clear. Until now, however, no single product has successfully addressed each of the identified customer issues of scalability, reliability, integration, and price/performance. After two years of preliminary study, followed by three years of research and development, data-FAST has achieved this goal.

The data-FAST product addresses these key customer issues as follows:

- *Scalability*. Virtually linear expandable growth in performance and capacity. Customers can now cost-effectively expand their DBMS applications from the smallest configuration to mainframe-class power within the context of a single, modular architecture.
- *Integration*. The DB FAST-MANAGER, which includes a server interface which supports SQL Server, slides seamlessly into existing SQL Server DBMS applications. No hardware or software is obsoleted and, most significantly, users notice only the improved performance and added storage capacity. Further, because the DB FAST-MANAGER is a dedicated server, other business computing requirements are isolated from and thus remain completely unaffected by the performance demands of the DBMS.
- *Reliability*. The distributed, parallel architecture of the product is inherently and significantly more reliable than past generations of DBMS solutions. Further, planned near-term enhancements will provide mainframe levels of fault tolerant reliability.

  The real significance and advantage of the data-FAST solution becomes apparent to corporate and company leaders when they analyze the economics of their present and future needs.
- *Price/Performance*. The data-FAST product sets a new standard for price/performance in the DBMS marketplace — across virtually the

entire performance spectrum of DBMS applications. No other current or anticipated DBMS product approaches the DB FAST-MANAGER level of price/performance.

Overview

The data-Fast DBMS server encompasses both the hardware and software required to perform the database server functions in a network computing environment, as shown in Figure 3 on p. 70. The DB FAST-MANAGER hardware features a modular multiprocessor architecture which allows linear growth in database processing power and disk storage. The DB FAST-MANAGER software makes optimum use of this multiprocessor architecture to divide the database storage and management functions among all available processors.

Multiprocessor Hardware Architecture

The multiprocessor architecture of the DB FAST-MANAGER, shown in Figure 4 on p. 71, allows modular expansion through the use of Database Processing Modules (DPMs) interconnected in a processor matrix. Each DPM consists of a high-performance 32-bit processor, memory, and SCSI disk drives. The DPMs are connected in a processor matrix via proprietary High-Speed Matrix Buses to form the DB FAST-MANAGER system.

DPMs plug into modular "stackbox" enclosures to configure the system of desired performance. The base configuration may consist of two DPMs plugged into a desktop or modular tower enclosure. As additional DPMs are added to the system, overall database performance (transactions per second, search speed, and capacity) increases linearly. The DB FAST-MANAGER can effectively grow from two to hundreds of DPMs as the customer's database performance and capacity requirements increase.

Each DB FAST-MANAGER system also incorporates one or more Interface Processor Modules (IPMs). The IPMs handle the interface with the outside world and the management of system backup devices, such as tape units and CD-ROMs. The IPMs manage the connection of DB FAST-MANAGER to one or more host computers or direct network connections, reference Figure 3. The IPMs accept SQL database transaction requests from the host machine or network and initiates their execution on the matrix of DPMs. The IPMs subsequently pass the transaction results back to the host system or network.

FIGURE 3: Distributed Client-Server

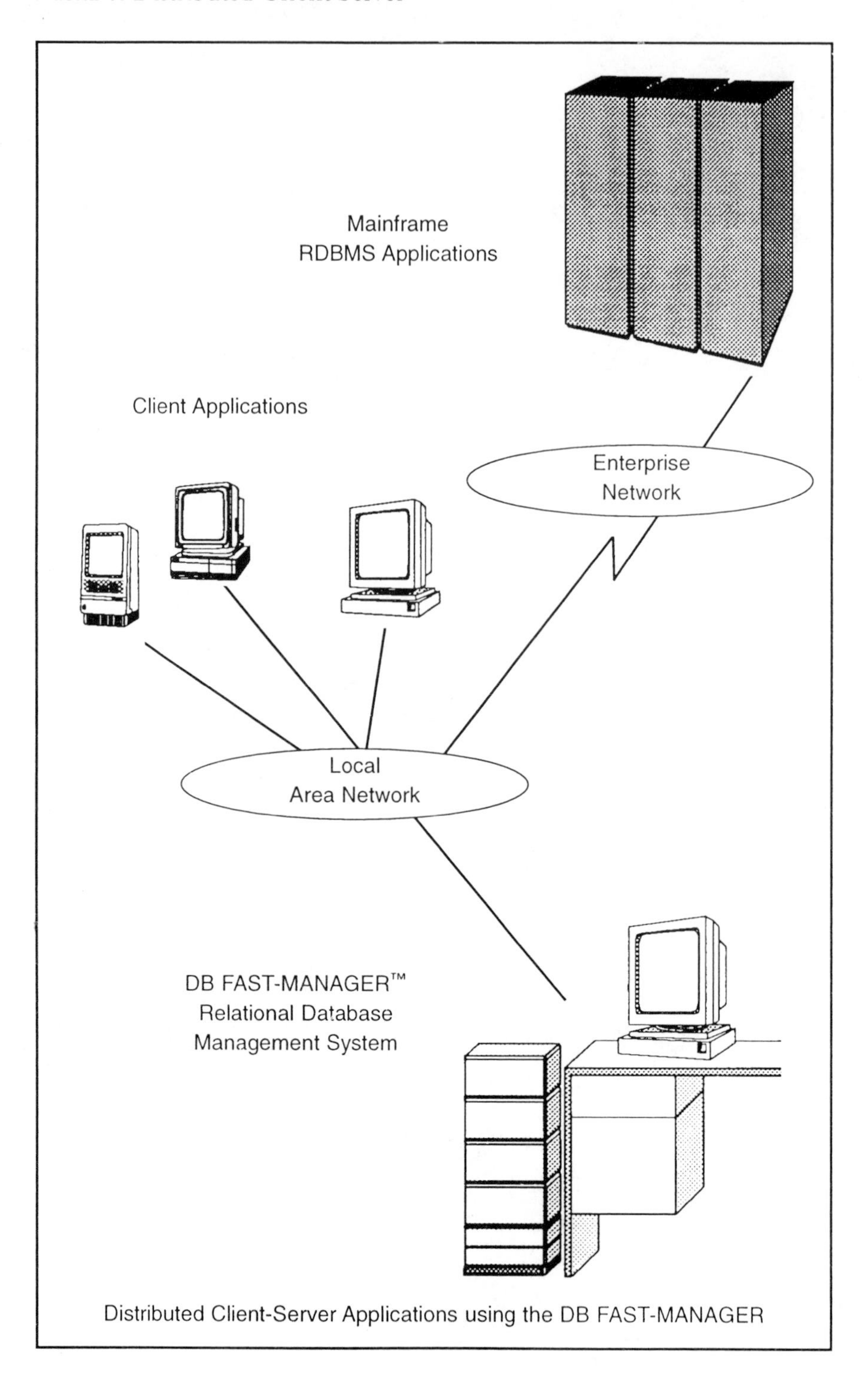

Distributed Client-Server Applications using the DB FAST-MANAGER

FIGURE 4: Multiprocessor Architecture

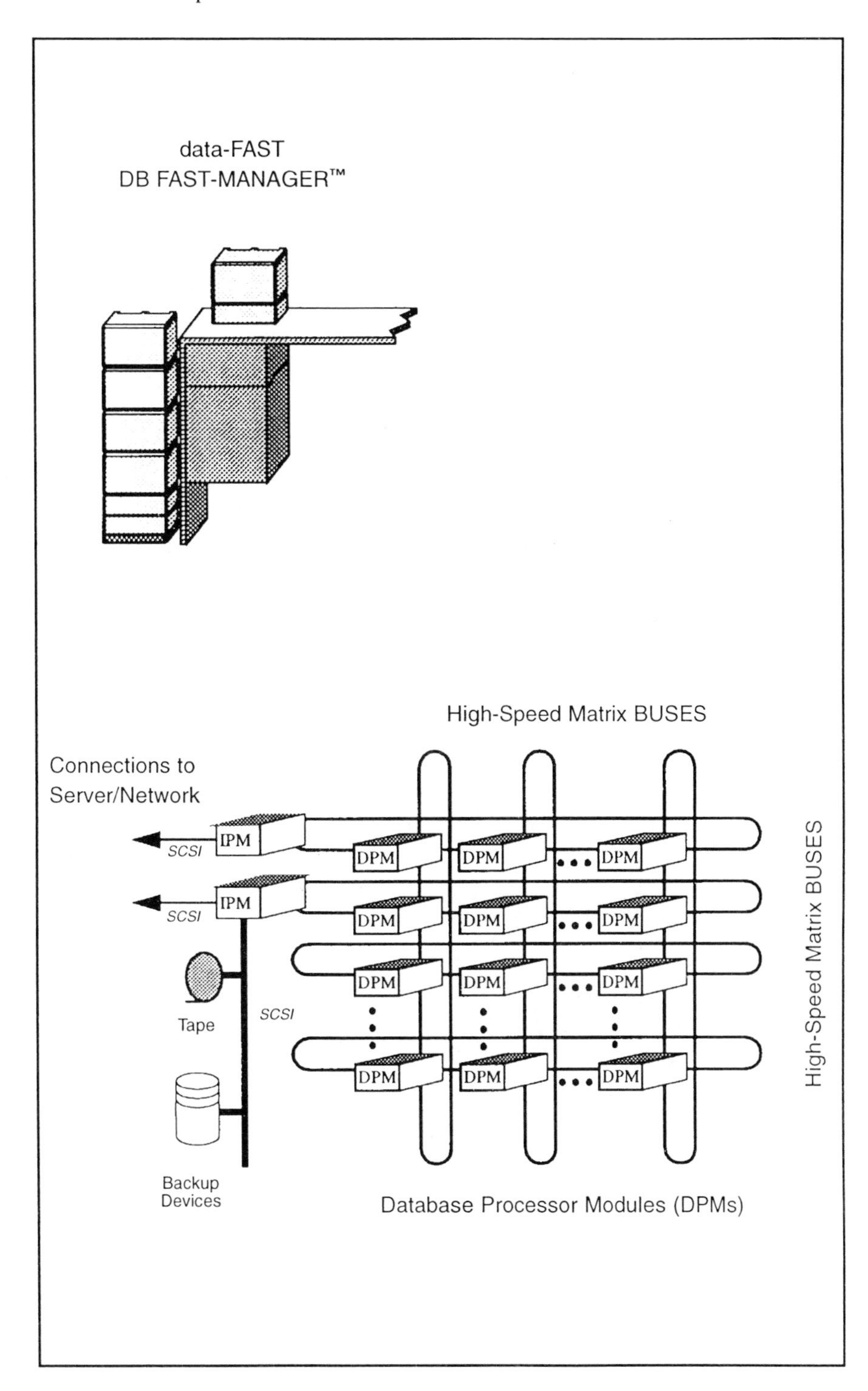

Integrated Relational Database Management System

Integral to the DB FAST-MANAGER is a full-featured Relational Database Management System (RDBMS) which makes optimal use of the system's multiprocessor architecture. When attached to a network server or directly on the network, the DB FAST-MANAGER provides the database processing function for multiple client workstations on the network, such as PCs and UNIX workstations. The DB FAST-MANAGER provides an ANSI SQL-2 compatible database management system. Compliance with the SQL standard allows integration with existing database systems and applications. In addition to SQL compatibility, data-FAST also delivers a compatible interface to SQL Server, the popular, commercially available RDBMS software. This allows the DB FAST-MANAGER to off-load the existing database applications.

The industry standard SQL command set is also augmented with transaction processing extensions to form DB FAST-SQL. DB FAST-SQL allows application developers to take maximum advantage of the DB FAST-MANAGER's transaction processing power.

The DB FAST-MANAGER's high performance is achieved through distributing relational databases and database operations over all of the DPMs. When a complex database search operation is performed — for example, find all invoices for red widgets sold to customers in New York in the last two years — all DPMs simultaneously work on the operation. In contrast, simple database transactions — for example, find the account balance for customer #x — are sent directly to the DPM, which manages that subset of the database. The proprietary software DBMS integrated in the DB FAST-MANAGER ensures an equal division of data and work which yields maximum performance.

Product Line Evolution

DB FAST-MANAGER Release 1. To be introduced in the fourth quarter of year one, consists of desktop and tower configurations of DB FAST-MANAGER oriented toward the small to mid-range RDBMS market segment. DB FAST-MANAGER Series SX-E2 is oriented towards database applications running on networks of PC and/or UNIX-based workstations. This product family is ideal for customers who are downsizing their database applications from large mainframe computers to networked, departmental systems.

DB FAST-MANAGER Release 2. The SX-ER system, to be introduced in the 3rd quarter of year two, will add fault-tolerance and additional hardware/software performance enhancements to the DB FAST-MANAGER line. This system will tolerate component hardware failures without loss of data or operational capabilities. The fault tolerant system will also provide features such as hot backups, mirrored data, and hot swap-out of DPMs. The fault tolerant DB FAST-MANAGER will allow data-FAST to sell into larger, more mission-critical database applications requiring hundreds of transactions per second, such as banking and finance. An integral host platform will also be developed.

DB FAST-MANAGER SX-E2 Series

The initial DB FAST-MANAGER SX-E2 Series consists of desktop and tower configurations of DB FAST-MANAGER oriented towards the small to mid-range DBMS market segment. These DB FAST-MANAGER models address database applications running on networks of PC and/or UNIX-based workstations. The SX-E2 models consist of both proprietary data-FAST SQL DBMS software and configured hardware modules. The system attaches to any host-server running UNIX or OS/2 on a network to off-load the existing relational database server, reference Figure 3 (p. 70).

The SX-E2D is a desktop model of the DB FAST-MANAGER which can accommodate up to four DPMs. The SX-E2D is available in a variety of configurations, from two DPMs to four DPMs, 128MB of memory, and 8.0 GB of disk storage. The SX-E2D provides a very competitively priced entry-level system into the DB FAST-MANAGER line.

The SX-E2T model provides a modular tower/rack design which allows from two to hundreds of DPMs within a modular stackbox architecture. An SX-E2T tower is configured from two DPMs to a total capacity of 12 DPMs, 384MB of memory, and 48GB of disk storage.

The DB FAST-MANAGER attaches to any UNIX or OS/2 server via a Small Computer Systems Interface (SCSI). A set of software modules running on the host-server, known as Host Database Services (HDS), handles all communication with the DB FAST-MANAGER, over the SCSI connection, reference Figure 4 (p. 71). The DB FAST-MANAGER includes a server interface which supports SQL Server. This allows the DB FAST-MANAGER to act as a plug-compatible database server for customers using applications written for SQL Server. Another advantage of the SQL Server interface is that it allows compatibility with a variety of database application development tools and front ends created by various vendors for SQL Server.

Because of the designed-in expandability and performance of the DB FAST-MANAGER system, it can easily grow to fill the needs of networks with hundreds of workstations and users. In addition, because it is a modular add-on product, all the customers' existing equipment and associated investments will be retained.

Product and Marketing Positioning

Obviously an emerging company with limited resources, a new product, and a new company name cannot immediately capture a marketplace. This business plan section describes how the product will be logically introduced into selected marketplaces over a scheduled time period.

In addition, since existing company products will heavily influence how the marketplace will view and accept your product, a Competitive Analysis is provided in this section. As in the marketing section, a thorough competition research study should be conducted and the results must be incontestably accurate.

This positioning section should immediately precede the sales section, as it describes the methodology which the company will use to achieve its sales goals.

IV. Product and Marketing Positioning

Overview

The marketplace for DB FAST-MANAGER is enormous, dynamic, and transitional. The landscape of companies span the spectrum from low-end PC system solutions to Fortune 1000 organizations with multi-million dollar mainframe class machines. Obviously, data-FAST cannot have it all and has identified specific market segments as described below.

The data-FAST product positioning strategy is formulated to take advantage of substantial shortcomings in current DBMS solutions and leverage the market's inherent volatility due to the unforeseen growth in database size and processing power. The strategy is two-phased and begins with direct selling to a carefully selected prospect profile to establish market presence and demand pull before broader markets are addressed.

Phase 1

The DBMS marketplace is currently weakest in its inability to offer smooth, cost-effective and seamless upgrades for users in need of higher performance and/or greater capacity. In particular, our primary focus is the rapidly growing base of UNIX platforms with DBMS resident software that have reached the upper limits of processing and/or capacity. Data-FAST offers current users a simple, elegant, and highly cost-effective strategy to off-load the DBMS from their existing general purpose UNIX platform onto a data-FAST DB FAST-MANAGER. This strategy frees up the computing resources and delays indefinitely the purchase of a more powerful computer. The turnkey installation of DB FAST-MANAGER is so easy and seamless that end users will continue to use their known structure and notice only an improvement in performance. Customers see DB FAST-MANAGER as a highly cost-effective, dedicated DBMS solution.

Phase 2

Upon successful implementation of phase one, DB FAST-MANAGER will have the name recognition and market presence to create demand pull from the user and vendor community. Data-FAST can position DB FAST-MANAGER as a specialized alternative platform of choice for DBMS users, Independent Software Vendors (ISVs) and value added resellers (VARs). In this phase the data-FAST sales organization will also begin to form alliances and original equipment manufacturer (OEM) relationships; increasing in time as potential end users demand that DB FAST-MANAGER be part of the vendors overall product offering. This strategy harvests the good name and product advantages built in phase one and moves DB FAST-MANAGER into the DBMS market mainstream.

In the accompanying chart (Figure 5 on p. 78), the market positioning strategy for the data-FAST product is depicted graphically vis-à-vis the competitive landscape. The Phase 1 marketing strategy moves customer's existing DBMS applications from their costly (both in dollars and performance) general purpose UNIX and OS/2-DOS platforms to the DB FAST-MANAGER.

In Phase 2, the data-FAST product will be positioned to attack the entire spectrum of market applications and competitors.

Competitive Analysis

As in any large and lucrative market there are many competitive alternatives for the user's budgeted funds. The two-phase positioning strategy is planned to minimize the competitive exposure of the data-FAST sales organization during the initial sales cycles. Therefore, the competitive picture is also seen in two phases.

Phase 1

No existing product approaches the market from DB FAST-MANAGER's unique position. However, our segmented and targeted audience for this phase will have alternatives to solve their problem of increasing performance and/or capacity. The prospect may choose to do one of the following:

FIGURE 5: Database Management System — Competitive Marketplace

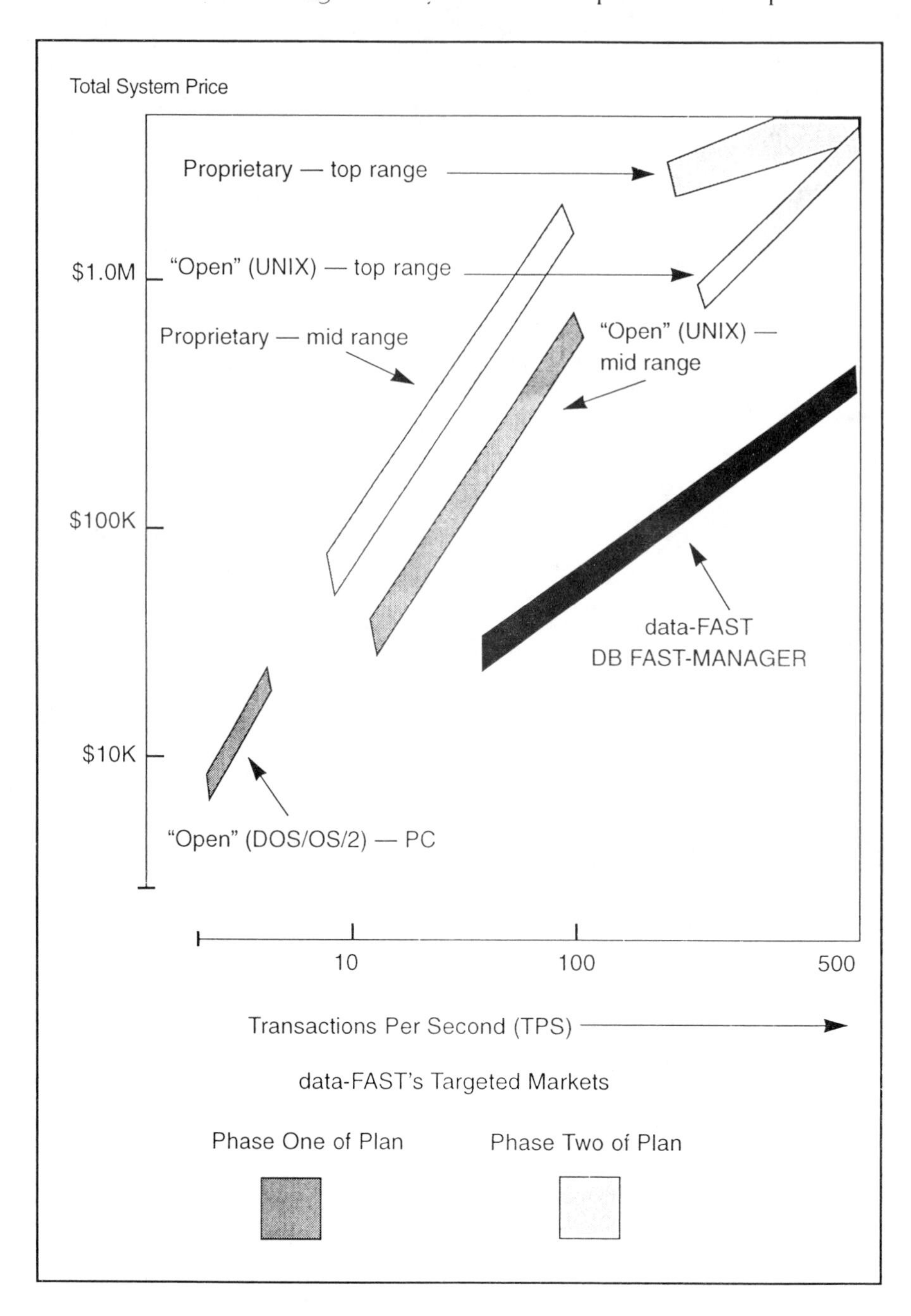

(a) Move up to a higher performance UNIX machine, buy a different general purpose UNIX machine, or consider a multiprocessing proprietary platform. All of these options are much more costly than option B.

(b) Buy a data-FAST DB FAST-MANAGER as a specialized DBMS applications solution and retain the existing UNIX platform as an open server or for other applications.

Phase 2

Now DB FAST-MANAGER is positioned as a general purpose solution for all DBMS environments and, as such, will become a candidate for new first-time applications.

To address the competitive environment in both phases, a complete Competitive Summary is provided in Figure 6 on pp. 80–81.

The price/performance of a data-FAST DB FAST-MANAGER is unsurpassed in the industry. DBMS software vendors rely on existing hardware vendors to provide upward scalability for their users. DB FAST-MANAGER provides true linear scalability [when the number of DPMs are doubled, system performance, measured in transactions per second (TPS), doubles]. DB FAST-MANAGER system tests have proven this to be true. Figure 7 on p. 82 shows the results of these tests with various numbers of DPMs installed.

FIGURE 6: Competitive Summary

	PC-Based Servers with DBMS Software	General Purpose Multi-User Computers w/DBMS Software
Description	Encompasses PC-based servers on a Local Area Network (LAN) running a DBMS software package. Client PCs (DOS, OS/2, Macs) access the DBMS databases on the LAN using a variety of database front-ends.	Large range of multi-user systems running DBMS software. These systems will be connected to a LAN and/or have multiple direct-connected terminals to allow users access to the databases. These systems are generally divided into Proprietary (proprietary H/W and operating system) or UNIX (running standard UNIX OS).
Hardware Components	386/486 Servers, single or dual-processor	UNIX systems: Sun, IBM, DEC, HP, NCR Proprietary Systems: IBM AS/400, DEC VAX, HP
Software Components	SQL Server, Gupta SQLBase, Oracle for OS/2, XDB server, DataEase, Novell NetWare SQL (running on DOS, OS/2, and NetWare)	Oracle DBMS (40% market-share), Informix Online, Sybase SQL Server, INGRES, DEC Rdb, IBM OS/400 DBMS, HP AllBase, etc.
System Price Range	H/W: $5,000–$30,000 S/W: $1,200–$12,000	H/W: $30,000–$2.5 million S/W: $12,000–$250,000
Typical System Price	10 users: H/W = $12,000 S/W = $2,495	15-20 users: H/W = $100,000 S/W = $20,000
Performance Range • TPS (TPC-B)	10–100 TPS	20–250 TPS
Price/TPS	$2,500–$5,000/TPS	$2,500–$30,000/TPS
Strengths	• Hardware is inexpensive, widely available, and easy to buy • Many customers already have a PC-based fileserver (just add software...) • Lower cost-per-TPS vs. large systems • Lots of front-ends, tools, and utilities available	• Have large install base and many existing applications • Hardware is general purpose • Strong sales and support (IBM,DEC,HP, Sun) • Lower systems costs vs. large mainframes • UNIX systems are touted as open
Weaknesses	• Growth: Customers have limited performance growth potential, (TPS, capacity) • Reliability: PC systems don't have all reliability features of large systems	• Growth and expandability in performance: Often customers must buy another computer in order to upgrade performance • Growth is costly and non-scalable • Cost/TPS is typically in the $5,000/TPS range
data-FAST **Competitive Position**	• Scalable Growth: Customers can grow from 20 TPS/0.8 GB to 2400 TPS/3.84 GB • data-FAST cost-per-TPS is competitive with PC-based systems • Reliability: data-FAST fault-tolerance, ECC memory	• Scalable Growth: Customers can grow from 20 TPS/0.8 GB to 2400 TPS/3.84 GB without ever throwing away any hardware • data-FAST cost/TPS is as low as 1/10th of these systems • DB FAST-MANAGER is an add-on to UNIX systems to off-load the DBMS

FIGURE 6: Competitive Summary, continued

	Special Purpose Database Machines	Multiprocessor UNIX Systems with DBMS Software
Description	Systems specifically designed as database machines. These systems focus on DBMS performance advantage through use of specialized hardware and software. These systems are usually attached to some host computer which serves as the database repository for a network of users attached via a LAN or direct connection.	Latest multiprocessor computer systems running multiprocessor versions of UNIX. DBMS software such as Oracle must be modified/rewritten to take advantage of these systems. These parallel systems are being touted as providing scalable growth in the high-end computer marketplace.
Hardware Components	Teradata DBC/1012 Charles River Relational Accelerator (Specific to Oracle on VAX)	Sun SPARCserver 600 MP, Sequent S2000 series, nCUBE, PARASYS, NCR
Software Components	DBMS software is proprietary and integrated with hardware. Focus is on plug-and-play with popular DBMS systems; TeraData to IBM DB2, Charles River to Oracle.	Oracle 6.2 recently announced to run on multiprocessor UNIX systems from Sequent, nCUBE, and Meiko. Sybase announced multiprocessor SQL server.
System Price Range	$100,000–$20 million for H/W & S/W	H/W: $80,000–$1 million S/W $10,000–$250,000
Typical System Price	Teradata: $1,000,000 Charles River: $80,000	40–60 users: H/W: $150,000 S/W: $50,000
Performance Range • TPS (TPC-B)	30–1000 TPS	30–1000 TPS
Price/TPS	$30,000/TPS	$2,500–$10,000/TPS
Strengths	• High performance; 100s of TPS • Price/performance: Better than mainframes $30k/TPS vs. $50 + k/TPS for mainframe • Reliability on par with mainframe systems • Compatible with existing DBMS software, (Teradata to IBM DB2, Charles River to Oracle)	• General-purpose UNIX computers running DBMS software follow trend to open systems • Price/performance: $5,000–$10,000/TPS vs. mainframes in the $50,000+/TPS range
Weaknesses	• No low-end and midrange solutions. System costs of $100,000+ are big-ticket items • Not touted as open systems • Non-scalable performance in the low-end and midrange DBMS segments	• Systems have high entry-level price in the $100K range. • Lack of software that takes advantage of multiprocessor hardware. Oracle and Sybase just announced, others coming. • H/W not optimal for database performance
data-FAST **Competitive Position**	• data-FAST provides scalable growth in the 20–2400 TPS range • data-FAST provides low entry-level solutions starting in the $20,000 range • data-FAST provides better price/performance at 1/10th the cost • data-FAST is compatible with popular DBMS systems	• data-FAST can position systems as add-ons to existing UNIX systems • data-FAST provides scalable growth in the 20–2400 TPS range • data-FAST provides low entry-level solutions starting in the $20,000 range • data-FAST optimized for DBMS performance: better price/performance • data-FAST is DBMS compatible

FIGURE 7: Price/Performance Comparison

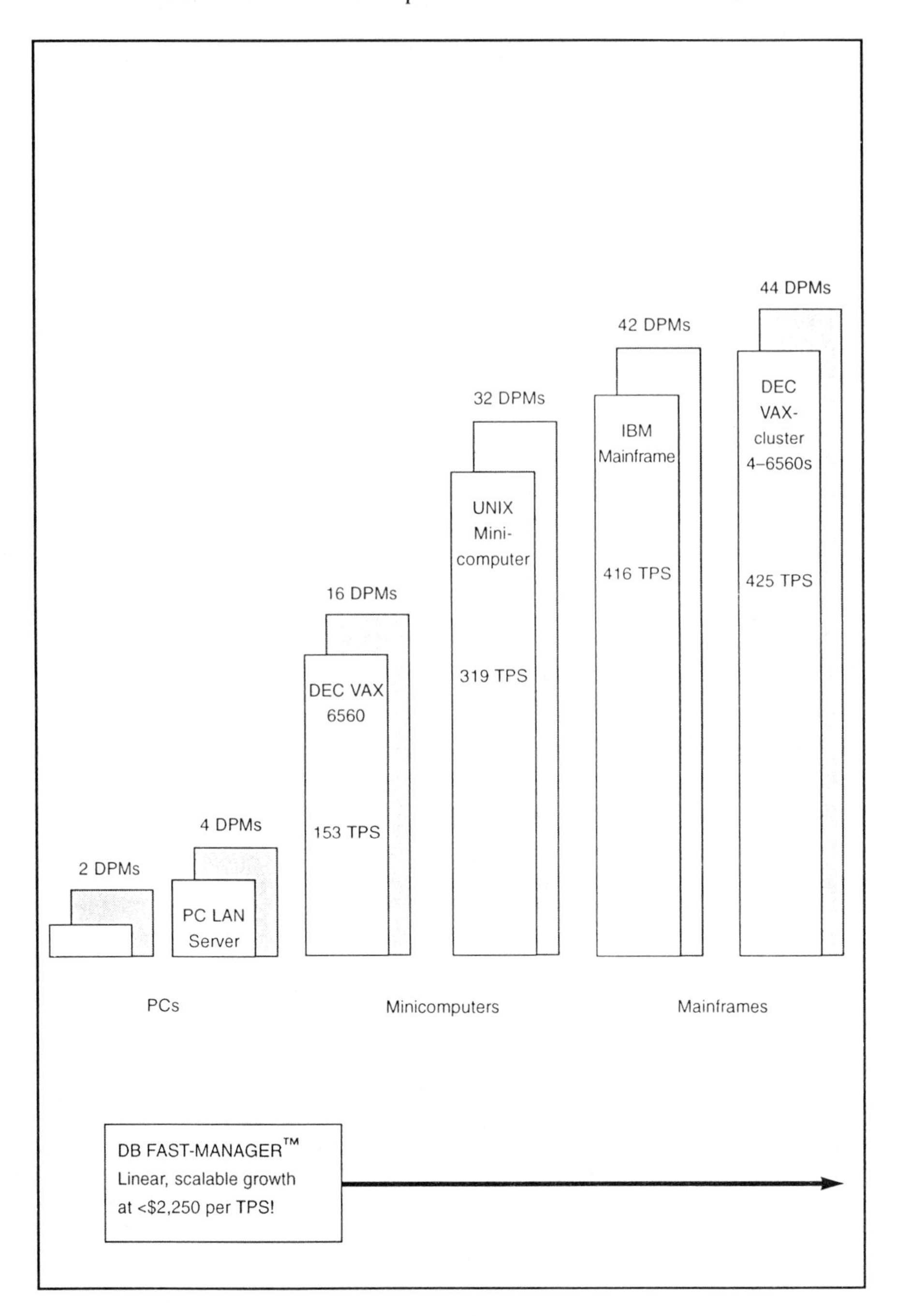

Sales

In the business plan's sales section, sales methodology is described and goals identified. Sales goals are believable if all the plan's other sections have, by now, convinced investors that you've created something that they may be interested in pursuing.

V. Sales Strategy and Forecast

Overview

Data-FAST is utilizing a direct sales organization to carry the message and introduce the DB FAST-MANAGER to the end user community. This direct channel will be expanded and continue to be our primary distribution channel throughout the second year. Our strategy is to choose carefully the initial installations to our best competitive advantage, thus building a credible reference list to leverage future efforts. Once references, credibility, and name recognition have been established, the sales organization will add Independent Software Vendors (ISVs) and Value Added Resellers (VARs) to its prospect lists. Alliances and strategic relationships with these accounts and later with original equipment manufacturers (OEMs) will give us additional leverage over and above what can be accomplished with our direct sales to end users.

Development of Direct Sales and Support Organization

The present direct sales organization for data-FAST consists of sales offices on each coast of the United States, Chicago, and in our headquarters' city, each supported by regional system engineers. These sales resources have had active prospects to call from a trade show data-FAST participated in earlier this year and from a list of current users that may need more power or capacity. These prospects are currently processing on UNIX machines and reaching the upper technical boundaries of their systems. Additional sales resources will be added throughout year two at the rate of one every other month. At least one support person will be hired per remote office and additional resources will be added as needed to maintain our excellent reputation for support.

Sales and support training classes are conducted at data-FAST headquarters, where students develop the appropriate skills to represent data-FAST products to their prospects. We will, of course, continue to spend time and resources on this important training process to ensure we stay current and accurate in this rapidly changing marketplace. Also, while certain variations in sales style are appropriate, it is important that all sales and support personnel present a common corporate theme.

The sales strategy theme is a modern approach to technical sales; a consultative sales approach that fosters a win-win, long-term relationship with the customer. This is important to data-FAST, for in the long run, as much as 50 percent of the company's growth will come from repeat business in upgrades and support. As we nurture and manage potential customers, we use the 8-4-2-1 milestones to measure our progress. That is, eight initial calls on qualified prospects produce four presentations; those four presentations produce two evaluations, and one of those evaluations becomes an order. By setting these standards and definitions, we facilitate communication among a geographically dispersed organization. This common language allows us to manage and control the sales cycle efficiently from long distances.

A lead tracking system has been implemented in the field offices using the 8-4-2-1 milestones. This system is then tied to a formal reporting document that tracks our progress as these leads become prospects and move toward evaluations and/or sales. The report consists of product description, revenue amounts, forecasted closing dates and a factored revenue total based on the accounts progress in the sales cycle. The sales teams report formally on a monthly basis and much more often on an ad hoc basis. With this information we can generate a 120-day rolling forecast. Our goal is to provide a revenue forecast that is within a 10 percent level of accuracy over the long term.

The technical support organization, made up of field and headquarters system engineers, is an integral part of the sales strategy and the sales team approach. While the level of product and environmental technical training is obviously higher for the system engineers; they are exposed to the same sales philosophies and techniques as the sales representatives. The whole team, therefore, can provide the best win-win environment for the customer. The field engineers are data-FAST's first line of attack. They are responsible for successful demonstrations, installations, and on-going customer support. They are backed up with a second level of support at headquarters, which is accessed via a hot-line. This hot-line is managed by headquarters system engineers who have access to anyone in the company to obtain the answers they need to serve the customer or sales effort.

International Sales

Once DB FAST-MANAGER is established in the U.S. and Canada, we will move into worldwide sales. Data-FAST has already begun negotiations with a trading firm to represent us in the Far East. We will bring representatives to our headquarters at the appropriate time, within the next six months, to test the product with their application software. Upon completion of testing and agreements, the training will be completed and prospecting can start.

At about the same time we will begin to develop a sales and support channel for Europe. Initial European contacts are being made to determine interest and alternatives. This channel will be a mixture of employees and agents (similar to that which the director of sales has arranged in the past) leveraging the best alternative in each country. We expect both of these areas to be contributing significantly to revenues in year three.

Development of Strategic Relationships

Once credibility has been established and good references are in place, the sales and support organization will turn some of their resources to establish strategic relationships with ISVs, VARs and OEMs. The primary goal of this strategy is to introduce DB FAST-MANAGER to existing sales networks and customer bases. These companies are the prime solution providers in the mid-range computing and networking environments. They provide turnkey solutions to application specific customer problems and are ideal conduits for data-FAST products. These relationships may take the form of reference sales where data-FAST is called into the account and we sell it. Or, we may be called into an account and joint market with the ISV or VAR. These markets include: financial services, insurance, education, health care, transportation, and utilities.

The ISVs and VARs will want to participate with data-FAST because we will make support for their products easier as their customers will no longer have to change hardware to grow. It will be easier for their customers to grow under DB FAST-MANAGER than any other alternative currently on the market.

Another step in the evolution of data-FAST's markets, and possibly one of the most significant, will be OEM or private label agreements with key international DBMS hardware/software vendors. The DBMS market will continue to come under pressure due to previously identified customer concerns such as linear scalability and price/performance. Industry leaders will seek out companies with products like DB FAST-MANAGER; products which can provide immediate solutions for their customers and reestablish the vendor's market position.

Data-FAST has evaluated the feasibility of such relationships and considers their formation likely. These relationships could very well include joint development projects where our product is tightly integrated into the vendor's solution and sold by the vendors sales and support group under their name. Or we could have the more typical OEM agreement where the vendor would sell DB FAST-MANAGER bundled with their solution touting our name and support. In any of the above situations, data-FAST would be leveraging the successful launch of the product line by our direct sales and support organization.

Figure 8 represents the projected sales forecast by the execution of this strategy through the plan period.

FIGURE 8: Sales Forecast — Plan Period

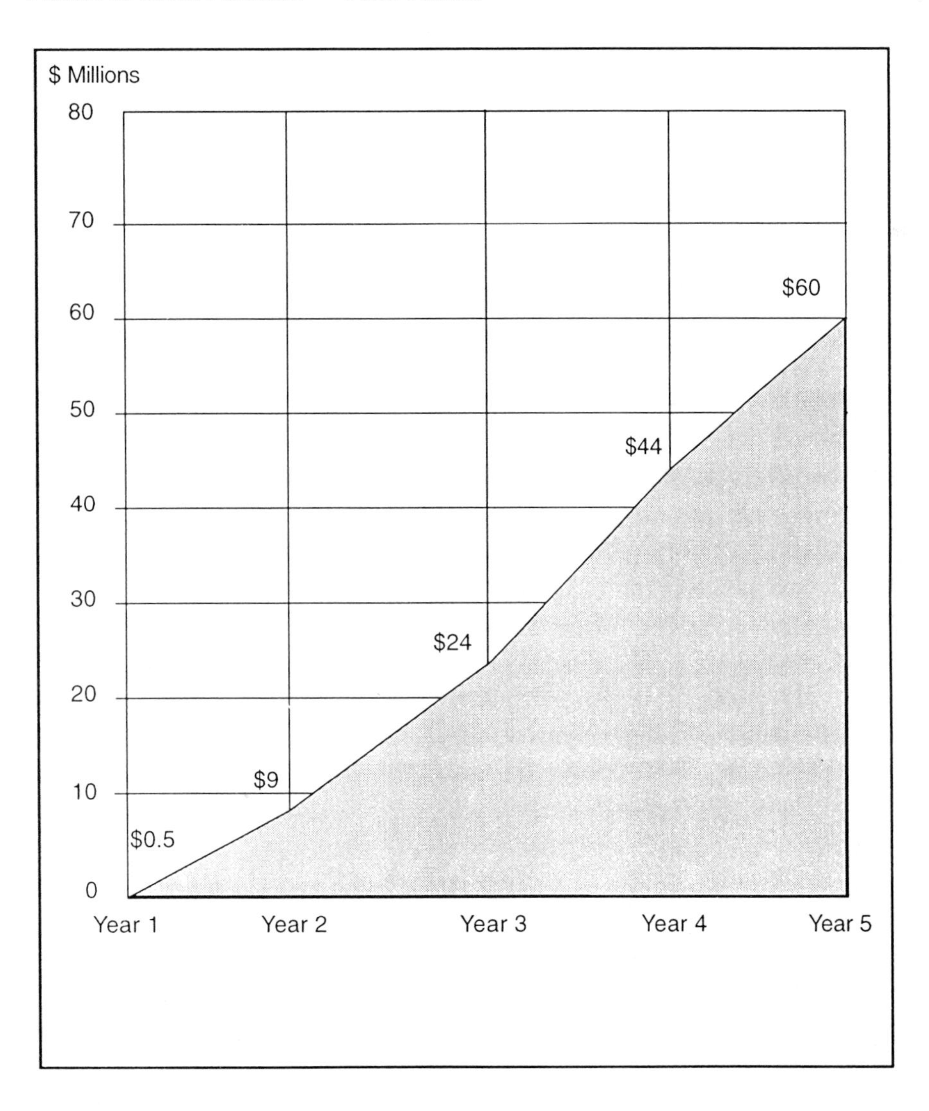

Operations

The business plan's operation section defines how manufacturing, product support, service, and quality control will be conducted.

VI. Operational Plan

Manufacturing

Data-FAST produces only the highest quality products. Our goal is to have 95 percent initial production yield with a 99 percent ongoing production yield. The data-FAST product is manufactured by qualified subcontractors, in order to reduce the development stage risks and minimize funding needs while ensuring high quality. These services are highly competitive and their use reduces the need for large capital outlays for manufacturing equipment in the early growth stages of the company. In addition subcontractors carry the cost of parts and finished goods. Subassemblies are fully tested with state-of-the-art automated test equipment by the supplier. Data-FAST evaluates a predetermined acceptable quality level to ensure ongoing quality. Subcontractors are evaluated monthly for quality and on-time delivery. Formal on-site evaluations are performed no less than quarterly. Manufacturing capabilities will be acquired in the future if it becomes advantageous to do so. Final system assembly, testing, and software integration is conducted at the company's facilities.

The manufacturing organization is staffed with individuals who are experienced in the planning and execution disciplines of manufacturing engineering, material planning and control, quality control, and large-scale production.

Product Support

Following the sale and installation of a data-FAST system, product support services are available via a 24-hour telephone service. Additionally, customers may subscribe to several levels and grades of support, ranging from on-site service calls, telephone aid, periodic software updates, and major new software releases. The 24-hour hot-line service for handling simple questions and/or to schedule service calls is included in the original purchase price. An active service contract is required to receive technical software support or on-site service or software updates.

Service Plan

Because the product design's inherent redundancy reduces failures the incidence of field service requests will also be reduced. In addition, the system's self-test diagnostics detect component failures and automatically redirect operations to redundant facilities. Furthermore, the company's local and remote software diagnostics rapidly detect system failures and reduce the mean time to repair. In-the-field replacements of failed elements are performed at the macro level only. Only major system components or subassemblies are exchanged in the event of a failure. In most cases, diagnosis and replacement are customer initiated, since the DP Module is the only required spare.

The company will maintain field service depots to inventory these major system components within close proximity to its regional sales offices. The company will also provide comprehensive field service training for its customers. An in-house staff of highly qualified technical support personnel will provide documentation, materials, and customer liaisons.

Production Costs

Based on present costs and through its own research and discussions with manufacturers, data-FAST has estimated its future projected cost of sales. Verbal price quotes have been obtained from these manufacturers in order to determine future production costs. The subcontractor is responsible for controlling material and labor costs. Their future estimates have been incorporated into the costs included in the financial projections.

Quality

As a corporate entity and as a group of individuals data-FAST is committed to cost-effectively producing the highest quality product. Adherence to the International Standards Organization (ISO) 9000 requirements enables three quality objectives: 1) An ability to achieve and sustain product quality that meets or exceeds the customer's stated or implied needs; 2) An ability to ensure organizational management that the specified level of quality is being achieved and sustained; 3) An ability to assure the customer that the requisite level of quality is or will be provided. Through compliance with ISO 9000 principles and the application of ISO 9004 quality management and system guidelines, data-FAST Corporation is actively preparing for ISO 9001 certification.

Financial

The business plan's financial section indicates what the company's present financial condition is and what it will be when its goals are met. The financial sheets are a result of all your hard work done in the previous sections.

Some investors will search through these sheets hoping to find an error — I suppose to put your management team on the defensive. Plugging in numbers and allowing spreadsheet packages to develop these financial sheets is OK; but, someone in the company should thoroughly understand all line items.

In presentation meetings, use these financial sheets to show how the investor's infusion of funds will increase the future value of the company and their resulting ROI.

Financial Plan Summary

Note: The following Financial Summary is inserted in the plan preceding the Assumptions.

"Management believes that total funding of $2,897,000, provided in two rounds, will carry it through the product marketing and company development phase, and into profitability. Start-up funds of $800,000 have been provided to complete the development of the system. The second round funding of $2,097,000 will provide the company with working capital while progressing to profitability in the second quarter of year two and into a positive cash flow position in the fourth quarter of year two."

Summary Chart

(In Thousands)	Year 1	Year 2	Year 3	Year 4	Year 5
Sales:	$481	$8,995	$24,000	$44,000	$60,000
Cost of Goods:	$286	$4,613	$11,869	$21,468	$29,160
Operating Expenses:	$1,227	$3,201	$7,883	$13,642	$17,286
Net Income:	($682)	$780	$2,804	$5,867	$8,946
Market Size:		$13 B	$15 B	$17 B	$21 B
Market Share:		0.07%	0.16%	0.25%	0.29%
Introductions of Product:	-a	-b			

Financial Assumptions

The following Assumptions' guidelines are inserted in the plan preceding the financial spreadsheets (Profit and Loss, Cash Flow, and Balance Sheets).

- Sales: The forecast assumes that sales of Product-a will commence by the 10th month; sales of Product-b by the 18th month.

  Indirect forecasted sales have been discounted by distributor costs of 44 percent.

- Cost of Sales: An eight percent reduction in the product cost in the second quarter of year two is assumed due to the following:

  - Material costs will decline, reflecting volume discounts due to increased levels of production.
  - Overhead will decline as fixed costs stabilize and sales volume increases.

Product costs will continue to decline, reflecting economies of scale and price reductions of components. The declining costs will offset the market squeeze caused by anticipated competition.

Direct sales assume a three percent commission.

- Marketing: The marketing program will commence in the second and third quarters of year one, six months before product shipments begin. Marketing expenses will increase sharply upon product introduction, reflecting associated selling and promotional costs. These costs eventually level off to 11 percent of sales.

- Research and Development: R&D costs will be very high during the prototype development and during various future product developments. These costs in year two are estimated at $1.4M and then taper off to 14 percent of sales by the fifth year.

- Receivables: Receivables will be collected 60 days following the product sale date.

- Payables: Outside services and product materials will be paid on a current basis in years one and two and within 30 days in years three through five. Five percent of operating expenses and all capital equipment will be paid within 30 days. All other vendors are paid upon invoice receipt.

  The Cash Flow Statement shows the change in payables only for operating activities while the Balance Sheet shows the balance of all payables (operations and capital equipment investments).

- Depreciation: Depreciation is calculated on all fixed and capital assets, assuming a useful life of five years and straight-line computation.

- Founders Investment: Founders have deferred existing expenses of $149,200, until the company is profitable.

- Short-Term Debt: No debt is assumed to be incurred in this plan.

- Long-Term Debt: A long-term capital lease line of $500,000 may be established in the third year to finance manufacturing equipment purchases. No debt is assumed to be incurred in this plan.

- Funding Needs: Working capital in the amount of $2,897,000 is needed to fund the conversion of the product to a consumer acceptable unit, and to fund the manufacturing and sales plan. The company anticipates that this infusion of funds will be sufficient to carry it to profitability.

- Capital Stock: The officers and employees will be afforded equity positions in the company during the start-up and as determined by the president. Stock allowances greater than five percent will require board approval.

Financial Spreadsheets

The Projected Income Statement, Cash Flow Statement, and Balance Spreadsheets are inserted in the plan following the Assumptions guidelines. The following spreadsheets are examples of how your financial sheets may appear.

Note: G & A stands for General and Administrative and R & D stands for Research and Development.

Projected Income Statement

Year 1
(in thousands of dollars)

	M01	M02	M03	M04	M05	M06
Sales	-	-	-	-	-	-
Less Cost of Goods	-	-	-	-	-	-
Gross Margin	-	-	-	-	-	-
Operating Expenses						
G & A	10	13	18	18	13	17
Marketing	-	1	1	15	16	22
R & D	9	22	35	40	50	41
Total	19	36	54	73	79	80
Pre-tax Income	(19)	(36)	(54)	(73)	(79)	(80)
Provision for Tax	(6)	(12)	(18)	(25)	(27)	(27)
Net Income	(13)	(24)	(36)	(48)	(52)	(53)

(Minor differences are due to rounding.)

Projected Income Statement, continued

Year 1
(in thousands of dollars)

	M07	M08	M09	M10	M11	M12	TOTAL
Sales	-	-	-	99	132	250	481
Less Cost of Goods	-	-	-	59	79	148	286
Gross Margin				40	53	102	195
Operating Expenses							
G & A	17	18	19	24	23	23	213
Marketing	29	30	30	132	92	102	470
R & D	55	52	54	54	64	68	544
Total	101	100	103	210	179	193	1227
Pre-tax Income	(101)	(100)	(103)	(170)	(126)	(91)	(1032)
Provision for Tax	(34)	(34)	(35)	(58)	(43)	(31)	(350)
Net Income	(67)	(66)	(68)	(112)	(83)	(60)	(682)

Projected Income Statement

Year 2
(in thousands of dollars)

	M01	M02	M03	M04	M05	M06
Sales	327	404	481	557	634	711
Less Cost of Goods	168	207	247	286	325	365
Gross Margin	159	197	234	271	309	346
Operating Expenses						
G & A	24	24	23	24	24	26
Marketing	86	86	95	95	100	153
R & D	84	98	99	98	98	99
Total	194	208	217	217	222	278
Pre-tax Income	(35)	(11)	17	54	87	68
Provision for Tax	(12)	(4)	6	18	30	23
Net Income	(23)	(7)	11	36	57	45

(Minor differences are due to rounding.)

Projected Income Statement, continued

Year 2
(in thousands of dollars)

	M07	M08	M09	M10	M11	M12	TOTAL
Sales	788	865	942	1019	1095	1172	8995
Less Cost of Goods	404	444	483	522	561	601	4613
Gross Margin	384	421	459	497	534	571	4382
Operating Expenses							
G & A	26	26	26	27	27	26	303
Marketing	124	114	124	142	121	125	1365
R & D	116	134	155	167	186	199	1533
Total	266	274	305	336	334	350	3201
Pre-tax Income	118	147	154	161	200	221	1181
Provision for Tax	40	50	52	55	68	75	401
Net Income	78	97	102	106	132	146	780

PROJECTED INCOME STATEMENT

Year 3
(in thousands of dollars)

	YEAR 3				
	Q01	**Q02**	**Q03**	**Q04**	**TOTAL**
Sales	4280	5430	6570	7720	24000
Less Cost of Goods	2117	2685	3249	3818	11869
Gross Margin	2163	2745	3321	3902	12131
Operating Expenses					
G & A	131	166	200	236	733
Marketing	541	687	831	977	3036
R & D	734	931	1126	1323	4114
Total	1406	1784	2157	2536	7883
Pre-tax Income	757	961	1164	1366	4248
Provision for Tax	257	327	396	464	1444
Net Income	500	634	768	902	2804

(Minor differences are due to rounding.)

Projected Income Statement

Years 4, 5
(in thousands of dollars)

			YEAR 4			YEAR 5
	Q01	Q02	Q03	Q04	TOTAL	TOTAL
Sales	9000	10330	11670	13000	44000	60000
Less Cost of Goods	4391	5040	5694	6343	21468	29160
Gross Margin	4609	5290	5976	6657	22532	30840
Operating Expenses						
G & A	273	313	354	394	1334	1806
Marketing	1067	1224	1383	1541	5215	6612
R & D	1451	1665	1881	2096	7093	886
Total	2791	3202	3618	4031	13642	17286
Pre-tax Income	1818	2088	2358	2626	8890	13554
Provision for Tax	618	710	802	893	3023	4608
Net Income	1200	1378	1556	1733	5867	8946

Cash Flow Statement

Year 1
(in thousands of dollars)

	M01	M02	M03	M04	M05	M06
Beginning Cash Bal.	0	27	30	43	36	30
Cash Receipts						
Accts. Receivable	-	-	-	-	-	-
Use of Cash						
Cost of Goods	-	-	-	-	-	-
Capital Equipment	10	52	13	33	5	16
Operating Expenses	15	31	66	55	78	66
Income Taxes	-	-	-	-	-	-
Total Disbursements	25	83	79	88	83	82
Sale of Stock	52	86	92	81	77	96
Ending Cash Balance	27	30	43	36	30	44

(Minor differences are due to rounding.)

Cash Flow Statement, continued

Year 1
(in thousands of dollars)

	M07	M08	M09	M10	M11	M12	TOTAL
Beginning Cash Bal.	44	0	0	(2)	1835	1550	0
Cash Receipts							
Accts. Receivable	-	-	-	-	-	99	99
Use of Cash							
Cost of Goods	-	-	-	59	79	148	286
Capital Equipment	14	29	15	-	30	80	297
Operating Expenses	103	102	99	201	176	187	1179
Income Taxes	-	-	-	-	-	-	-
Total Disbursements	117	131	114	260	285	415	1762
Sale of Stock	73	131	112	2097	-	-	2897
Ending Cash Balance	0	0	(2)	1835	1550	1234	1234

Cash Flow Statement

Year 2

(in thousands of dollars)

	M01	M02	M03	M04	M05	M06
Beginning Cash Bal.	1234	987	766	612	515	455
Cash Receipts						
Accts. Receivable	132	250	327	404	481	557
Use of Cash						
Cost of Goods	168	207	247	286	325	365
Capital Equipment	19	62	24	5	2	23
Operating Expenses	192	202	210	210	214	266
Income Taxes	-	-	-	-	-	-
Total Disbursements	379	471	481	501	541	654
Sale of Stock	-	-	-	-	-	-
Ending Cash Balance	987	766	612	515	455	358

(Minor differences are due to rounding.)

Cash Flow Statement, continued

Year 2

(in thousands of dollars)

	M07	M08	M09	M10	M11	M12	TOTAL
Beginning Cash Bal.	358	51	35	20	0	5	1234
Cash Receipts							
Accts. Receivable	634	711	788	865	942	1019	7110
Use of Cash							
Cost of Goods	404	444	483	522	561	601	4613
Capital Equipment	283	23	31	46	56	31	605
Operating Expenses	254	260	289	317	320	334	3068
Income Taxes	-	-	-	-	-	-	-
Total Disbursements	941	727	803	885	937	966	8286
Sale of Stock	-	-	-	-	-	-	-
Ending Cash Balance	51	35	20	0	5	58	58

Cash Flow Statement

Year 3
(in thousands of dollars)

	YEAR 3				
	Q01	**Q02**	**Q03**	**Q04**	**TOTAL**
Beginning Cash Bal.	58	670	688	849	58
Cash Receipts					
Accts. Receivable	3567	4660	5810	6960	20997
Use of Cash					
Cost of Goods	1425	2495	3063	3627	10610
Capital Equipment	134	178	181	188	681
Operating Expenses	1344	1712	2078	2446	7580
Income Taxes	52	257	327	396	1032
Total Disbursements	2955	4642	5649	6657	19903
Sale of Stock	-	-	-	-	-
Ending Cash Balance	670	688	849	1152	1152

(Minor differences are due to rounding.)

Cash Flow Statement

Years 4, 5
(in thousands of dollars)

			YEAR 4			YEAR 5
	Q01	**Q02**	**Q03**	**Q04**	**TOTAL**	**TOTAL**
Beginning Cash Bal.	1152	1715	2424	3298	1152	4373
Cash Receipts						
Accts. Receivable	8120	9450	10770	12120	40460	58010
Use of Cash						
Cost of Goods	4204	4823	5478	6132	20637	28847
Capital Equipment	196	209	212	213	830	466
Operating Expenses	2693	3091	3496	3898	13178	16695
Income Taxes	464	618	710	802	2594	4350
Total Disbursements	7557	8741	9896	11045	37239	50358
Sale of Stock	-	-	-	-	-	-
Ending Cash Balance	1715	2424	3298	4373	4373	12025

BALANCE SHEET

Year 1

(in thousands of dollars)

	M01	M02	M03	M04	M05
Assets					
Current Assets					
Cash	27	30	43	36	30
Accts. Receivable	-	-	-	-	-
Prepaid Expenses	0	1	1	1	1
Total Current Assets	27	31	44	37	31
Equipment Less					
Accumulated Depr.	10	61	72	103	106
Deferred Tax Asset	6	18	36	61	88
Total Assets	**43**	**110**	**152**	**201**	**225**
Liabilities					
Current Liabilities					
Accts. Payable	4	8	(6)	9	8
Income Taxes	-	-	-	-	-
Total Current Liabilities	4	8	(6)	9	8
Total Liabilities	**4**	**8**	**(6)**	**9**	**10**
Stockholder's Equity					
Capital Stock	52	138	230	311	388
Retained Earnings	(13)	(36)	(72)	(119)	(173)
Total Stockholder's Equity	39	102	158	192	215
Total Liabilities and Stockholder's Equity	**43**	**110**	**152**	**201**	**225**

Balance Sheet, continued

Year 1

(in thousands of dollars)

	M06	M07	M08	M09	M10	M11	Yr. End M12
Assets							
Current Assets							
Cash	44	0	0	(2)	1835	1550	1234
Accts. Receivable	-	-	-	-	99	231	382
Prepaid Expenses	1	1	0	0	0	0	0
Total Current Assets	45	1	0	(2)	1934	1781	1616
Equipment Less							
Accumulated Depr.	119	131	171	168	195	270	284
Deferred Tax Asset	115	149	183	218	276	319	350
Total Assets	**279**	**281**	**354**	**384**	**2405**	**2370**	**2250**
Liabilities							
Current Liabilities							
Accts. Payable	19	23	23	9	44	92	33
Income Taxes	-	-	-	-	-	-	-
Total Current Liabilities	19	23	23	9	44	92	33
Total Liabilities	**21**	**17**	**25**	**11**	**47**	**92**	**33**
Stockholder's Equity							
Capital Stock	484	557	688	800	2897	2897	2897
Retained Earnings	(226)	(293)	(359)	(427)	(539)	(619)	(680)
Total Stockholder's Equity	258	264	329	373	2358	2278	2217
Total Liabilities and Stockholder's Equity	**279**	**281**	**354**	**384**	**2405**	**2370**	**2250**

BALANCE SHEET

Year 2

(in thousands of dollars)

	M01	M02	M03	M04	M05
Assets					
Current Assets					
Cash	987	766	612	515	455
Accts. Receivable	577	731	885	1038	1191
Prepaid Expenses	1	1	1	1	1
Total Current Assets	1565	1498	1499	1556	1649
Equipment Less					
Accumulated Depr.	340	357	355	350	365
Deferred Tax Asset	362	366	360	342	312
Total Assets	**2267**	**2221**	**2214**	**2248**	**2326**
Liabilities					
Current Liabilities					
Accts. Payable	71	34	16	12	33
Income Taxes	-	-	-	-	-
Total Current Liabilities	71	34	16	12	33
Total Liabilities	**71**	**34**	**16**	**12**	**33**
Stockholder's Equity					
Capital Stock	2897	2897	2897	2897	2897
Retained Earnings	(701)	(710)	(699)	(661)	(604)
Total Stockholder's Equity	2196	2187	2198	2236	2293
Total Liabilities and Stockholder's Equity	**2267**	**2221**	**2214**	**2248**	**2326**

Balance Sheet, continued

Year 2

(in thousands of dollars)

	M06	M07	M08	M09	M10	M11	Yr. End M12
Assets							
Current Assets							
Cash	358	51	35	20	0	5	58
Accts. Receivable	1345	1499	1653	1807	1961	2114	2267
Prepaid Expenses	1	1	1	0	0	0	0
Total Current Assets	1707	1554	1693	1831	1961	2119	2325
Equipment Less							
Accumulated Depr.	640	650	667	704	745	760	770
Deferred Tax Asset	289	249	199	147	92	24	0
Total Assets	**2636**	**2453**	**2559**	**2682**	**2798**	**2903**	**3095**
Liabilities							
Current Liabilities							
Accts. Payable	296	36	44	65	73	47	42
Income Taxes	-	-	-	-	-	-	52
Total Current Liabilities	296	36	44	65	73	47	94
Total Liabilities	**296**	**36**	**44**	**65**	**73**	**47**	**94**
Stockholder's Equity							
Capital Stock	2897	2897	2897	2897	2897	2897	2897
Retained Earnings	(557)	(480)	(382)	(280)	(172)	(41)	104
Total Stockholder's Equity	2340	2417	2515	2617	2725	2856	3001
Total Liabilities and Stockholder's Equity	**2636**	**2453**	**2559**	**2682**	**2798**	**2903**	**3095**

BALANCE SHEET

Year 3

(in thousands of dollars)

	YEAR 3			
	Q01	**Q02**	**Q03**	**Q04**
Assets				
Current Assets				
Cash	670	688	849	1152
Accts. Receivable	2980	3750	4510	5270
Prepaid Expenses	1	1	1	0
Total Current Assets	3651	4439	5360	6422
Equipment Less Accumulated Depr.	877	997	1101	1211
Deferred Tax Asset	-	-	-	-
Total Assets	**4528**	**5436**	**6461**	**7633**
Liabilities				
Current Liabilities				
Accts. Payable	768	971	1160	1360
Income Taxes	257	327	396	464
Total Current Liabilities	1025	1298	1556	1824
Total Liabilities	**1025**	**1298**	**1556**	**1824**
Stockholder's Equity				
Capital Stock	2897	2897	2897	2897
Retained Earnings	606	1241	2008	2912
Total Stockholder's Equity	3503	4138	4905	5809
Total Liabilities and Stockholder's Equity	**4528**	**5436**	**6461**	**7633**

BALANCE SHEET

Years 4, 5

(in thousands of dollars)

	YEAR 4				YEAR 5
	Q01	Q02	Q03	Q04	
Assets					
Current Assets					
Cash	1715	2424	3298	4373	12024
Accts. Receivable	6150	7030	7930	8810	10800
Prepaid Expenses	2	2	1	1	2
Total Current Assets	7867	9456	11229	13184	22826
Equipment Less Accumulated Depr.	1315	1425	1521	1609	1450
Deferred Tax Asset	-	-	-	-	-
Total Assets	**9182**	**10881**	**12750**	**14793**	**24276**
Liabilities					
Current Liabilities					
Accts. Payable	1552	1780	2001	2218	2494
Income Taxes	618	710	802	893	1152
Total Current Liabilities	2170	2490	2803	3111	3646
Total Liabilities	**2170**	**2490**	**2803**	**3111**	**3646**
Stockholder's Equity					
Capital Stock	2897	2897	2897	2897	2897
Retained Earnings	4115	5494	7050	8785	17733
Total Stockholder's Equity	7012	8391	9947	11682	20630
Total Liabilities and Stockholder's Equity	**9182**	**10881**	**12750**	**14793**	**24276**

Glossary

The final section of the business plan is a courtesy glossary to aid non-technical readers. Include this section in your business plans to help alleviate the investor's "I don't understand this ...." concern — you may not always be present to explain.

Glossary

American National Standards Institute (ANSI): An organization devoted to the development of voluntary standards that will enhance the productivity and international competitiveness of American industrial enterprises.

back-end: The portion of a program that does not interact with the user and that accomplishes the processing job that the program is designed to perform. In a local area network, the back-end programs handle the user interface on each workstation.

benchmarks: A variety of widely accepted standard tests you can run to try to compare the performance of various servers. Most benchmarks applied to DBMS are oriented to transaction processing (TP) and yield results in transactions/second (TPS). TP1 or TPC Benchmarks, which models debit/credit transactions for a banking environment, is the benchmark you'll see referred to most often. Expect results between 10 and 100 transactions per second (TPS) with micro servers. You may also read about Dewitt or Palmer benchmarks, which emphasize queries rather than on-line transaction processing (See OLTP).

client: In a local area network, a workstation with processing capabilities, such as a personal computer, that can request information or applications from the network's file server.

client-based application: In a local area network, an application that resides on a personal computer workstation and is not available for use by others on the network. Client-based applications do not make sharing common data easy, but they are resistant to the system-wide failure that occurs when a server-based application becomes unavailable after the file server crashes.

client-server architecture (CSA): A design model for applications running on a local area network, in which the bulk of the back-end processing, such as performing a physical search of a database, takes place on the file server.

The front-end processing, which involves communicating with the user, is handled by smaller programs distributed to the client workstations.

client-server network: A method of allocating resources in a local area network so that computing power is distributed among the personal computers in the network, but some shared resources are centralized in a file server.

database management system (DBMS): 1. In mainframe computing, a computer system organized for the systematic management of a large collection of information. 2. In personal computing, a program such as dBASE with similar information storage, organization, and retrieval capacities, sometimes including simultaneous access to multiple databases through a shared file.

database query: A request for information from a database expressed in a high-level language. Special software analyzes the user's request and responds with the information requested. In general usage the user's request may also update the database.

database server: A special type of back-end (see back-end) that provides high performance database services to front-end (see front-end) applications, usually in a network environment. The server is responsible for storing data and guaranteeing its integrity, processing queries, and handling security.

Digital Equipment Corporation (DEC): produces the popular VAX line of minicomputers. VAX/VMS is the most widely used operating system of VAX computers. The command language is similar to MS-DOS.

distributed processing: A superset of client-server architecture (CSA). Users and their applications have "transparent" access to data that might be spread across a variety of different computers situated in geographically different locations. Data and/or processing could theoretically be distributed. Named services need to be incorporated into LANs before distributed database processing can be a simple "mix-and-match" affair.

distributed computing systems: In a distributed system, the object is to get computing power to the user without giving up the means to share external computing resources, such as access to a central database. An example of a distributed computing system is a network of professional workstations.

downsizing: Rewriting mainframe or minicomputer-based applications to run on PCs.

fault tolerance: The ideal situation whereby hardware, software, and data never fail or become corrupted. This is typically accomplished via a combination of hardware and data redundancy, and special programming to control the redundancy. Disk mirroring and disk duplexing are two forms of fault tolerance.

front-end: The portion of a program that interacts directly with the user. In a local area network, this portion of the program may be distributed to each workstation so that the user can interact with the back-end application on the file server.

host: In a computer network, the computer that performs centralized functions such as making program or data files available to workstations in the network.

LAN: An acronym for Local Area Network. Hardware and software systems that undertake the job of inter-device communications within limited distance.

mainframe computer: A large, expensive computer generally used for information processing in large businesses, colleges, and organizations. Originally, the phrase referred to the extensive array of large rack and panel cabinets that enclosed early computers.

microcomputer: The smallest and least expensive class of computers. They are fully operational computers that use microprocessors as their CPU. Microcomputers include most popular personal computers and workstations.

minicomputer: A computer that is usually more powerful than a microcomputer, and usually less powerful than a mainframe computer.

multicomputer: A computer system that utilizes more than one CPU but does not have a shared central memory.

multiprocessor: A computer system consisting of two or more central processors having either a common memory or common control. In general usage, however, it also includes multicomputers.

Online Transaction Processing (OLTP): Describes real-time transaction processing. Associated terms: mission critical, benchmarks.

Operating System/2 (OS/2): A multitasking operating system for IBM PC-compatible computers that provides protection for programs running simultaneously, and enables the dynamic exchange of data between applications. IBM and Microsoft Corporation jointly developed and introduced OS/2 in 1987; they designed OS/2 as a replacement for MS-DOS.

PC: An abbreviation for Personal Computer.

platform (hardware): A computer hardware standard, such as IBM PC-compatible or Macintosh personal computers, in which a comprehensive approach to the computer solution of a problem can be based.

parallel processing: What a computer does when it carries out more than one computation at the same time on different CPUs.

Relational Database Management System (RDBMS): A database management program utilizing an organized method of information storage and retrieval, especially one that comes with all the necessary support programs and documentation needed to create, install, and maintain custom database applications.

scalability: The ability to use the same software environment on a broad spectrum of hardware. Generally considered a desirable feature of software; it tends to support growth and portability with a minimum of disruption.

server: A component of a computer network that provides information processing services. These services may include computation, display, or printing but data storage and retrieval is the most common.

SQL: An acronym for Structured Query Language which is an ANSI standard relational database query language (see ANSI).

TPS: Transactions per second.

transaction: A unit of work for a database system which corresponds to one debit or credit transaction in a banking environment. This has become one standard measure for performance comparisons among database systems.

UNIX: An operating system for a wide variety of computers, from mainframes to personal computers, that supports multitasking and is ideally suited to multiuser applications. Written in the highly portable programming language C, UNIX (like C) is the product of work at AT&T Bell Laboratories during the early 1970s. Originally developed by highly advanced research scientists for sophisticated work in computer science.

VAR: Value added reseller.

workstation: A configuration of computer equipment designed for use by one person at a time. Workstations are often connected to a computer network which provides other information services to the workstation.

Worksheets

The following worksheets are included for your use. The page numbers refer to the main text pages (all located under Management Leader Tasks in Chapter III), which will aid you in composing and filling in these sheets.

- Sales Forecast Chart (see Sales write-up in the Financial Section pp. 13–14)
- Summary Chart (see Financial Plan Summary write-up p. 19)
- Return on Investment (ROI) Analysis (see ROI Analysis pp. 20–22)
- Projected Income Statement (see write-up in the Financial Section p. 17)
- Cash Flow Statement (see Cash Flow Statement write-up in the Financial Section pp. 17–18)
- Balance Sheet (see Balance Sheet write-up in the Financial Section pp. 18–19)

Sales Forecast Chart

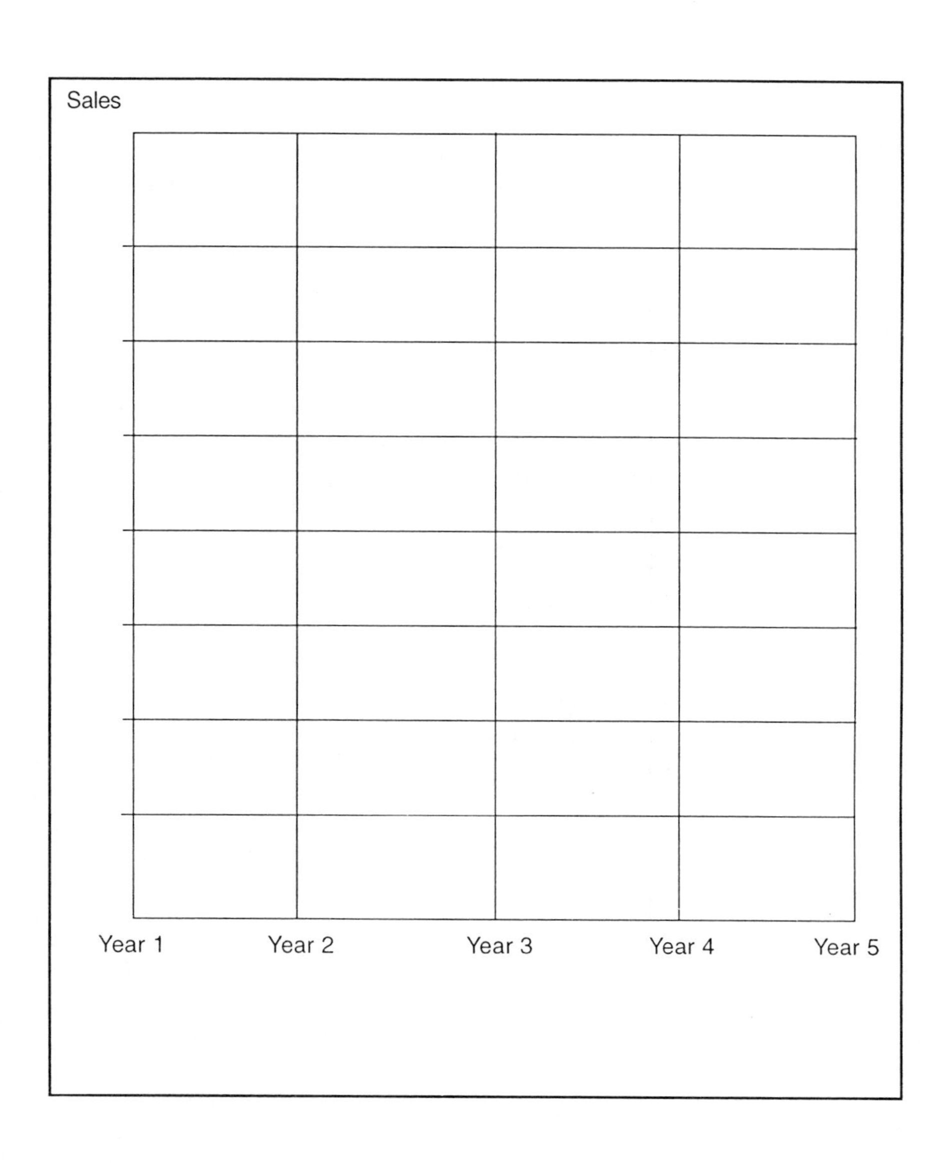

Summary Chart

(In Thousands)	Year 1	Year 2	Year 3	Year 4	Year 5
Sales:					
Cost of Goods:					
Operating Expenses:					
Net Income:					
Market Size:					
Market Share:					
Introductions of Product:					

Return on Investment (ROI) Analysis

Present Value Method: assuming a ten times investor return in five years:

	$M
Sales in year 5	
Investment needed	
5th year profits	
Company valuation in year 5 = (P/E) x (5th year profits) =	
Present company value = (company valuation) ÷ $(1+i)^n$ =	
i = 58% (10 times return), n = 5 (in 5 years)	
Investor's company share (Investment) ÷ (present value) =	%

Replication Method: assuming a ten times investor return in five years:

	MONTHS
Concept phase	
Study and product definition phase	
Product and company development phase	
Total Months (TM)	

	$M
Annualized profits of your company at TM	
Company valuation at TM = (P/E) x (annualized profits at TM) =	
Present company value = (company valuation at TM) ÷ $(1+i)^n$ =	
i = 58% (10 times return), n = TM ÷ 12	
Investor's company share (Investment) ÷ (present value) =	%

Projected Income Statement

Year 1
(in thousands of dollars)

	M01	M02	M03	M04	M05	M06
Sales						
Less Cost of Goods						
Gross Margin						
Operating Expenses						
G & A						
Marketing						
R & D						
Total						
Pre-tax Income						
Provision for Tax						
Net Income						

(Minor differences are due to rounding.)

Projected Income Statement, continued

Year 1
(in thousands of dollars)

	M07	M08	M09	M10	M11	M12	TOTAL
Sales							
Less Cost of Goods							
Gross Margin							
Operating Expenses							
G & A							
Marketing							
R & D							
Total							
Pre-tax Income							
Provision for Tax							
Net Income							

Projected Income Statement

Year 2

(in thousands of dollars)

	M01	M02	M03	M04	M05	M06
Sales						
Less Cost of Goods						
Gross Margin						
Operating Expenses						
G & A						
Marketing						
R & D						
Total						
Pre-tax Income						
Provision for Tax						
Net Income						

(Minor differences are due to rounding.)

Projected Income Statement, continued

Year 2
(in thousands of dollars)

	M07	M08	M09	M10	M11	M12	TOTAL
Sales							
Less Cost of Goods							
Gross Margin							
Operating Expenses							
G & A							
Marketing							
R & D							
Total							
Pre-tax Income							
Provision for Tax							
Net Income							

Projected Income Statement

Year 3
(in thousands of dollars)

			YEAR 3		
	Q01	**Q02**	**Q03**	**Q04**	**TOTAL**
Sales					
Less Cost of Goods					
Gross Margin					
Operating Expenses					
G & A					
Marketing					
R & D					
Total					
Pre-tax Income					
Provision for Tax					
Net Income					

(Minor differences are due to rounding.)

Projected Income Statement

Years 4, 5
(in thousands of dollars)

			YEAR 4			YEAR 5
	Q01	Q02	Q03	Q04	TOTAL	TOTAL
Sales						
Less Cost of Goods						
Gross Margin						
Operating Expenses						
G & A						
Marketing						
R & D						
Total						
Pre-tax Income						
Provision for Tax						
Net Income						

Cash Flow Statement

Year 1
(in thousands of dollars)

	M01	M02	M03	M04	M05	M06
Beginning Cash Bal.						
Cash Receipts						
Accts. Receivable						
Use of Cash						
Cost of Goods						
Capital Equipment						
Operating Expenses						
Income Taxes						
Total Disbursements						
Sale of Stock						
Ending Cash Balance						

(Minor differences are due to rounding.)

Cash Flow Statement, continued

Year 1
(in thousands of dollars)

	M07	M08	M09	M10	M11	M12	TOTAL
Beginning Cash Bal.							
Cash Receipts							
Accts. Receivable							
Use of Cash							
Cost of Goods							
Capital Equipment							
Operating Expenses							
Income Taxes							
Total Disbursements							
Sale of Stock							
Ending Cash Balance							

Cash Flow Statement

Year 2
(in thousands of dollars)

	M01	M02	M03	M04	M05	M06
Beginning Cash Bal.						
Cash Receipts						
Accts. Receivable						
Use of Cash						
Cost of Goods						
Capital Equipment						
Operating Expenses						
Income Taxes						
Total Disbursements						
Sale of Stock						
Ending Cash Balance						

(Minor differences are due to rounding.)

Cash Flow Statement, continued

Year 2
(in thousands of dollars)

	M07	M08	M09	M10	M11	M12	TOTAL
Beginning Cash Bal.							
Cash Receipts							
Accts. Receivable							
Use of Cash							
Cost of Goods							
Capital Equipment							
Operating Expenses							
Income Taxes							
Total Disbursements							
Sale of Stock							
Ending Cash Balance							

Cash Flow Statement

Year 3
(in thousands of dollars)

	YEAR 3				
	Q01	Q02	Q03	Q04	TOTAL
Beginning Cash Bal.					
Cash Receipts					
Accts. Receivable					
Use of Cash					
Cost of Goods					
Capital Equipment					
Operating Expenses					
Income Taxes					
Total Disbursements					
Sale of Stock					
Ending Cash Balance					

(Minor differences are due to rounding.)

Cash Flow Statement

Years 4, 5
(in thousands of dollars)

			YEAR 4			YEAR 5
	Q01	Q02	Q03	Q04	TOTAL	TOTAL
Beginning Cash Bal.						
Cash Receipts						
Accts. Receivable						
Use of Cash						
Cost of Goods						
Capital Equipment						
Operating Expenses						
Income Taxes						
Total Disbursements						
Sale of Stock						
Ending Cash Balance						

Balance Sheet

Year 1

(in thousands of dollars)

	M01	M02	M03	M04	M05
Assets					
Current Assets					
Cash					
Accts. Receivable					
Prepaid Expenses					
Total Current Assets					
Equipment Less					
Accumulated Depr.					
Deferred Tax Asset					
Total Assets					
Liabilities					
Current Liabilities					
Accts. Payable					
Income Taxes					
Total Current Liabilities					
Total Liabilities					
Stockholder's Equity					
Capital Stock					
Retained Earnings					
Total Stockholder's Equity					
Total Liabilities and Stockholder's Equity					

Balance Sheet, continued
Year 1
(in thousands of dollars)

	M06	M07	M08	M09	M10	M11	Yr. End M12
Assets							
Current Assets							
Cash							
Accts. Receivable							
Prepaid Expenses							
Total Current Assets							
Equipment Less Accumulated Depr.							
Deferred Tax Asset							
Total Assets							
Liabilities							
Current Liabilities							
Accts. Payable							
Income Taxes							
Total Current Liabilities							
Total Liabilities							
Stockholder's Equity							
Capital Stock							
Retained Earnings							
Total Stockholder's Equity							
Total Liabilities and Stockholder's Equity							

Balance Sheet

Year 2

(in thousands of dollars)

	M01	M02	M03	M04	M05
Assets					
Current Assets					
Cash					
Accts. Receivable					
Prepaid Expenses					
Total Current Assets					
Equipment Less					
Accumulated Depr.					
Deferred Tax Asset					
Total Assets					
Liabilities					
Current Liabilities					
Accts. Payable					
Income Taxes					
Total Current Liabilities					
Total Liabilities					
Stockholder's Equity					
Capital Stock					
Retained Earnings					
Total Stockholder's Equity					
Total Liabilities and Stockholder's Equity					

BALANCE SHEET, CONTINUED

Year 2

(in thousands of dollars)

	M06	M07	M08	M09	M10	M11	Yr. End M12
Assets							
Current Assets							
Cash							
Accts. Receivable							
Prepaid Expenses							
Total Current Assets							
Equipment Less Accumulated Depr.							
Deferred Tax Asset							
Total Assets							
Liabilities							
Current Liabilities							
Accts. Payable							
Income Taxes							
Total Current Liabilities							
Total Liabilities							
Stockholder's Equity							
Capital Stock							
Retained Earnings							
Total Stockholder's Equity							
Total Liabilities and Stockholder's Equity							

BALANCE SHEET

Year 3

(in thousands of dollars)

	YEAR 3			
	Q01	Q02	Q03	Q04
Assets				
Current Assets				
Cash				
Accts. Receivable				
Prepaid Expenses				
Total Current Assets				
Equipment Less				
Accumulated Depr.				
Deferred Tax Asset				
Total Assets				
Liabilities				
Current Liabilities				
Accts. Payable				
Income Taxes				
Total Current Liabilities				
Total Liabilities				
Stockholder's Equity				
Capital Stock				
Retained Earnings				
Total Stockholder's Equity				
Total Liabilities and Stockholder's Equity				

Balance Sheet

Years 4, 5

(in thousands of dollars)

	YEAR 4				YEAR 5
	Q01	Q02	Q03	Q04	
Assets					
Current Assets					
Cash					
Accts. Receivable					
Prepaid Expenses					
Total Current Assets					
Equipment Less					
Accumulated Depr.					
Deferred Tax Asset					
Total Assets					
Liabilities					
Current Liabilities					
Accts. Payable					
Income Taxes					
Total Current Liabilities					
Total Liabilities					
Stockholder's Equity					
Capital Stock					
Retained Earnings					
Total Stockholder's Equity					
Total Liabilities and Stockholder's Equity					

Index

A

accounts receivable, 17, 18
acquisitions, 30
administration and finance dept., 16
apportionment, 40
asset, deferred tax, 19
assets, 18
  capital, 94
  combined, 6
  partners', 5
  total, 19
assumptions: document, 18
assumptions, financial. *See* financial assumptions

B

balance sheet, 11, 13, 17, 18, 19
board of directors, 41
budgets, departmental, 27
business plan, 9, 25, 47

C

capital equipment, 18, 27
  requirements, 11, 18
  worksheet, 19
capital stock, 19, 94
capital, working, 94
cash flow, 13, 43
  manager, 44
  statement, 17, 19, 94
cash flow forecast, 11
collateral, 39
competitive analysis, 11
competitive summary, 79
competitor section, 24
confidential information, 34, 35
cost of goods, 27
cost of goods sold, 11, 15, 18
costs, ownership, 18
covenants, 39
cover and title page, 25

D

debentures, 39
debt, 94
deferred tax asset, 19
departments,
  administration, 16
  manufacturing, 17; Technical; 17
depreciation, 94

E

engineering and product 27

equity partnership, 8
equity, percentage of, 40
executive summary, 10, 13, 53
exhibits, 39
expenses,
prepaid, 18
operating, 11, 15, 16, 18

F

field engineers, 85
finance,
department, 16
phasing, 41
financial,
assumptions, 11, 16, 17, 18, 19, 93
plan, 27
plan summary, 19, 92
reports, 27
section, 13, 91
sheets, 13, 15, 20
statements, 13
summary, 11, 57
summary sheet, 33
financial,
leader, 12
performance, 56
projections, 16
statements warranty, 38
financing, 47
description, 38
staged, 41
founders investment, 94
funding, 3, 4, 19
future, 42
milestones, 41
needs, 94
sources, 47
structure, 40
funds needed, 22

G

general partner, 7, 8

I

income,
statement, 11
taxes, 18, 19
informal equity investors, 6
ISO (International Sales Organization) 9000, 90
investing partner's resource, 6
investment opportunity, 11, 19, 20, 33, 56
investment partners, 44
investor,
cash, 43
ownership, 14, 21
risk, 41
investor's company share, 21
investor's equity interest, 14
investor's ROI, 62
investors, early-on, 23
investment opportunity, 11, 19, 20, 33, 56
investment partners, 44

J

joint operating arrangement, 7

K

key management, 56

L

lawyer involvement, 38
liabilities, 19
limited partnerships, 30

M

management, 10, 13
cash, 43
key, 56
money, 44
people, 44
profile, 26
management leader, 2, 12, 16, 17, 32, 40, 43
management team, 2, 6, 12, 40, 60, 91
manufacturing, 89
concept, 27
costs, 15, 17
department, 17
engineering, 89
organization, 89
market,
analysis, 65
development, 20, 40

expansion, 46
niche, 23, 24
overview, 53
positioning, 26
presence, 77
share, 19
size, 19, 24, 33
marketing, 10, 27, 46, 93
department, 16
leader, 2, 12, 22
position, 10
schedules, 24
studies, 2
tasks, 23
marketplace, 13, 24, 26, 40, 62
needs, 46
opportunity, 19
mergers, 30
milestones, 42
funding, 41; 8-4-2-1, 85

N

negotiations, 37, 38
nondisclosure agreement, 34
nondisclosure statement, 36

O

objective, 3
company, 22
operating, control, 41
costs, 17
expenses, 11, 15, 16, 18
operational plan, 27, 89
operations, 11, 88
ownership,
percentages, 21, 42
position, 40

P

P/E ratios, 20, 33
partnership agreement, 7
partnership, limited, 30
payables, 19, 94
personnel, 15
management, 22
other, 61
profile, 3
requirements, 11, 15
private placements, 29
product 46, 67
description, 10
development schedules, 14
development stage, 41
line evolution, 72
marketability, 29
marketing positioning, 26, 75, 76
overview, 54
support, 89
production costs, 90
projected income, 13
statement, 17
property, 18

Q

quality, 90

R

receivables, 94
replication ROI analysis, 21
representations, 38
research & development (R&D), 90
research & development (R&D) venture, 4, 7
R&D Partnership (RDP), 7
R&D expenditures, 16
return on investment (ROI), 45, 91
analysis, 11, 14, 19, 20
calculations, 33
investors, 62
present value, 21
replication, 21
revenue,
amounts, 85
factored, 85
forecast, 85
royalty license, 7

S

sale of stock, 18
sales, 10, 13, 27, 83, 93
actual, 43
direct, 84, 93
international, 85

methodology, 83
philosophies, 85
projected, 20, 43
strategy, 27, 85\
strategy and forecast, 84
team, 85
sales forecast chart, 14, 87
securities, resale, 39
service plan, 90
statements,
cash flow, 17, 18, 19
projected income, 17, 19
stock,
capital, 19, 94
class of, 39
conversion privileges, 39
purchase warrant agreement, 40
stockholder's equity, 19
subordination, 39
summary, 25, 26, 29, 53–58
financial plan, 19
synergistic partner, 4, 7
partnerships, 30
teams, 44
system engineers, 85

T

technical department, 17
leader 2, 23
See technology leader
technical support, 90
organization, 85
technology leader, 22
license, 7

V

venture capitalists (V/Cs), 5, 6, 29

W

warranties, 38
warranty, financial statement, 38
worksheet, capital equipment, 19
worksheets, 13

ORDER DIRECTLY FROM THE OASIS PRESS®

THE OASIS PRESS® ORDER FORM

Call, Mail, Email, or Fax Your Order to: PSI Research, 300 North Valley Drive, Grants Pass, OR 97526 USA
Email: psi2@magick.net *Website:* http://www.psi-research.com
Order Phone USA & Canada: +1 800 228-2275 *Inquiries & International Orders:* +1 541 479-9464 *Fax:* +1 541 476-1479

TITLE	✔ BINDER	✔ PAPERBACK	QUANTITY	COST
Bottom Line Basics	❑ $39.95	❑ $19.95		
The Business Environmental Handbook	❑ $39.95	❑ $19.95		
Business Owner's Guide to Accounting & Bookkeeping		❑ $19.95		
Buyer's Guide to Business Insurance	❑ $39.95	❑ $19.95		
Collection Techniques for a Small Business	❑ $39.95	❑ $19.95		
A Company Policy and Personnel Workbook	❑ $49.95	❑ $29.95		
Company Relocation Handbook	❑ $39.95	❑ $19.95		
CompControl: The Secrets of Reducing Worker's Compensation Costs	❑ $39.95	❑ $19.95		
Complete Book of Business Forms		❑ $19.95		
Customer Engineering: Cutting Edge Selling Strategies	❑ $39.95	❑ $19.95		
Develop & Market Your Creative Ideas		❑ $15.95		
Doing Business in Russia		❑ $19.95		
Draw The Line: A Sexual Harassment Free Workplace		❑ $17.95		
The Essential Corporation Handbook		❑ $21.95		
The Essential Limited Liability Company Handbook	❑ $39.95	❑ $21.95		
Export Now: A Guide for Small Business	❑ $39.95	❑ $24.95		
Financial Management Techniques for Small Business	❑ $39.95	❑ $19.95		
Financing Your Small Business		❑ $19.95		
Franchise Bible: How to Buy a Franchise or Franchise Your Own Business	❑ $39.95	❑ $24.95		
Friendship Marketing: Growing Your Business by Cultivating Strategic Relationships		❑ $18.95		
Home Business Made Easy		❑ $19.95		
Incorporating Without A Lawyer (Available for 32 states) SPECIFY STATE:		❑ $24.95		
Joysticks, Blinking Lights and Thrills		❑ $18.95		
The Insider's Guide to Small Business Loans	❑ $29.95	❑ $19.95		
InstaCorp – Incorporate In Any State (Book & Software)		❑ $29.95		
Keeping Score: An Inside Look at Sports Marketing		❑ $18.95		
Know Your Market: How to Do Low-Cost Market Research	❑ $39.95	❑ $19.95		
Legal Expense Defense: How to Control Your Business' Legal Costs and Problems	❑ $39.95	❑ $19.95		
Location, Location, Location: How to Select the Best Site for Your Business		❑ $19.95		
Mail Order Legal Guide	❑ $45.00	❑ $29.95		
Managing People: A Practical Guide		❑ $21.95		
Marketing Mastery: Your Seven Step Guide to Success	❑ $39.95	❑ $19.95		
The Money Connection: Where and How to Apply for Business Loans and Venture Capital	❑ $39.95	❑ $24.95		
People Investment	❑ $39.95	❑ $19.95		
Power Marketing for Small Business	❑ $39.95	❑ $19.95		
Profit Power: 101 Pointers to Give Your Business a Competitive Edge		❑ $19.95		
Proposal Development: How to Respond and Win the Bid	❑ $39.95	❑ $21.95		
Raising Capital	❑ $39.95	❑ $19.95		
Retail in Detail: How to Start and Manage a Small Retail Business		❑ $15.95		
Secrets to Buying and Selling a Business		❑ $24.95		
Secure Your Future: Financial Planning at Any Age	❑ $39.95	❑ $19.95		
The Small Business Insider's Guide to Bankers		❑ $18.95		
Start Your Business *(Available as a book and disk package – see back)*		❑ $ 9.95 (without disk)		
Starting and Operating a Business in...series *Includes FEDERAL section PLUS ONE STATE section*	❑ $34.95	❑ $27.95		
PLEASE SPECIFY WHICH STATE(S) YOU WANT:				
STATE SECTION ONLY (BINDER NOT INCLUDED) SPECIFY STATE(S):	❑ $8.95			
FEDERAL SECTION ONLY (BINDER NOT INCLUDED)	❑ $12.95			
U.S. EDITION (FEDERAL SECTION – 50 STATES AND WASHINGTON DC IN 11-BINDER SET)	❑ $295.95			
Successful Business Plan: Secrets & Strategies	❑ $49.95	❑ $27.95		
Successful Network Marketing for The 21st Century		❑ $15.95		
Surviving and Prospering in a Business Partnership	❑ $39.95	❑ $19.95		
TargetSmart! Database Marketing for the Small Business		❑ $19.95		
Top Tax Saving Ideas for Today's Small Business		❑ $16.95		
Which Business? Help in Selecting Your New Venture		❑ $18.95		
Write Your Own Business Contracts	❑ $39.95	❑ $24.95		
BOOK SUB-TOTAL (FIGURE YOUR TOTAL AMOUNT ON THE OTHER SIDE)				

OASIS SOFTWARE Please check Macintosh or 3-1/2" Disk for IBM-PC & Compatibles

TITLE	3-1/2" IBM Disk	Mac-OS	Price	QUANTITY	COST
California Corporation Formation Package ASCII Software	☐	☐	$ 39.95		
Company Policy & Personnel Software Text Files	☐	☐	$ 49.95		
Financial Management Techniques (Full Standalone)	☐		$ 99.95		
Financial Templates	☐	☐	$ 69.95		
The Insurance Assistant Software (Full Standalone)	☐		$ 29.95		
Start A Business (Full Standalone)	☐		$ 49.95		
Start Your Business (Software for Windows™)	☐		$ 19.95		
Successful Business Plan (Software for Windows™)	☐		$ 99.95		
Successful Business Plan Templates	☐	☐	$ 69.95		
The Survey Genie - Customer Edition (Full Standalone)	☐		$149.95		
The Survey Genie - Employee Edition (Full Standalone)	☐		$149.95		
SOFTWARE SUB-TOTAL					

BOOK & DISK PACKAGES Please check whether you use Macintosh or 3-1/2" Disk for IBM-PC & Compatibles

TITLE	IBM-PC	Mac-OS	BINDER	PAPERBACK	QUANTITY	COST
The Buyer's Guide to Business Insurance w/ Insurance Assistant	☐		☐$ 59.95	☐$ 39.95		
California Corporation Formation Binder Book & ASCII Software	☐	☐	☐$ 69.95	☐$ 59.95		
Company Policy & Personnel Book & Software Text Files	☐	☐	☐$ 89.95	☐$ 69.95		
Financial Management Techniques Book & Software	☐		☐$ 129.95	☐$ 119.95		
Start Your Business Paperback & Software (Software for Windows™)	☐			☐$ 24.95		
Successful Business Plan Book & Software for Windows™	☐		☐ $125.95	☐ $109.95		
Successful Business Plan Book & Software Templates	☐	☐	☐ $109.95	☐$ 89.95		
BOOK & DISK PACKAGE TOTAL						

AUDIO CASSETTES

TITLE	Price	QUANTITY	COST
Power Marketing Tools For Small Business	☐ $ 49.95		
The Secrets To Buying & Selling A Business	☐ $ 49.95		
AUDIO CASSETTE SUB-TOTAL			

OASIS SUCCESS KITS Call for more information about these products

TITLE	Price	QUANTITY	COST
Start-Up Success Kit	☐ $ 39.95		
Business At Home Success Kit	☐ $ 39.95		
Financial Management Success Kit	☐ $ 44.95		
Personnel Success Kit	☐ $ 44.95		
Marketing Success Kit	☐ $ 44.95		
OASIS SUCCESS KITS TOTAL			

COMBINED SUB-TOTAL (FROM THIS SIDE)

SOLD TO: *Please give street address*

NAME:

Title:

Company:

Street Address:

City/State/Zip:

Daytime Phone: Email:

SHIP TO: *If different than above, please give alternate street address*

NAME:

Title:

Company:

Street Address:

City/State/Zip:

Daytime Phone:

YOUR GRAND TOTAL

SUB-TOTALS (from other side) $

SUB-TOTALS (from this side) $

SHIPPING (see chart below) $

TOTAL ORDER $

If your purchase is:	Shipping costs within the USA:
$0 - $25	$5.00
$25.01 - $50	$6.00
$50.01 - $100	$7.00
$100.01 - $175	$9.00
$175.01 - $250	$13.00
$250.01 - $500	$18.00
$500.01+	4% of total merchandise

PAYMENT INFORMATION: *Rush service is available, call for details.*
International and Canadian Orders: Please call for quote on shipping.

☐ CHECK Enclosed payable to PSI Research   Charge: ☐ VISA ☐ MASTERCARD ☐ AMEX ☐ DISCOVER

Card Number: Expires:

Signature: Name On Card:

High Tech 10/97

Call toll free to order 1-800-228-2275 PSI Research 300 North Valley Drive, Grants Pass, OR 97526 FAX 541-476-1479

Use this form to register for an advance notification of updates, new books and software releases, plus special customer discounts!

Please answer these questions to let us know how our products are working for you, and what we could do to serve you better.

Funding High-Tech Ventures

Rate this product's overall quality of information:
☐ Excellent
☐ Good
☐ Fair
☐ Poor

Rate the quality of printed materials:
☐ Excellent
☐ Good
☐ Fair
☐ Poor

Rate the format:
☐ Excellent
☐ Good
☐ Fair
☐ Poor

Did the product provide what you needed?
☐ Yes ☐ No

If not, what should be added?
____

This product is:
☐ Clear and easy to follow
☐ Too complicated
☐ Too elementary

Were the worksheets easy to use?
☐ Yes ☐ No ☐ N/A

Should we include?
☐ More worksheets
☐ Fewer worksheets
☐ No worksheets

How do you feel about the price?
☐ Lower than expected
☐ About right
☐ Too expensive

How many employees are in your company?
☐ Under 10 employees
☐ 10 - 50 employees
☐ 51 - 99 employees
☐ 100 - 250 employees
☐ Over 250 employees

How many people in the city your company is in?
☐ 50,000 - 100,000
☐ 100,000 - 500,000
☐ 500,000 - 1,000,000
☐ Over 1,000,000
☐ Rural (Under 50,000)

What is your type of business?
☐ Retail
☐ Service
☐ Government
☐ Manufacturing
☐ Distributor
☐ Education

What types of products or services do you sell?
____

What is your position in the company?
(please check one)
☐ Owner
☐ Administrative
☐ Sales/Marketing
☐ Finance
☐ Human Resources
☐ Production
☐ Operations
☐ Computer/MIS

How did you learn about this product?
☐ Recommended by a friend
☐ Used in a seminar or class
☐ Have used other PSI products
☐ Received a mailing
☐ Saw in bookstore
☐ Saw in library
☐ Saw review in:
  ☐ Newspaper
  ☐ Magazine
  ☐ Radio/TV

Where did you buy this product?
☐ Catalog
☐ Bookstore
☐ Office supply
☐ Consultant

Would you purchase other business tools from us?
☐ Yes ☐ No

If so, which products interest you?
☐ EXECARDS® Communications Cards
☐ Books for business
☐ Software

Would you recommend this product to a friend?
☐ Yes ☐ No

Do you use a personal computer?
☐ Yes ☐ No

If yes, which?
☐ Macintosh
☐ PC Compatible
☐ Other

Check all the ways you use computers?
☐ Word processing
☐ Accounting
☐ Spreadsheet
☐ Inventory
☐ Order processing
☐ Design/Graphics
☐ General Data Base
☐ Customer Information
☐ Scheduling
☐ Internet

May we call you to follow up on your comments?
☐ Yes ☐ No

May we add your name to our mailing list? ☐ Yes ☐ No

If you'd like us to send associates or friends a catalog, just list names and addresses on back.

Is there anything we should do to improve our products?
____

Just fill in your name and address here, fold (see back) and mail.

Name ____________________
Title ____________________
Company ____________________
Phone ____________________
Address ____________________
City/State/Zip ____________________
Email Address (Home) __________ (Business) __________

If you have friends or associates who might appreciate receiving our catalogs, please list here. Thanks!

Name________________________ Name________________________
Title________________________ Title________________________
Company________________________ Company________________________
Phone________________________ Phone________________________
Address________________________ Address________________________
Address________________________ Address________________________

FOLD HERE FIRST

NO POSTAGE
NECESSARY
IF MAILED
IN THE
UNITED STATES

BUSINESS REPLY MAIL

FIRST CLASS MAIL PERMIT NO. 002 MERLIN, OREGON

POSTAGE WILL BE PAID BY ADDRESSEE

PSI Research
PO BOX 1414
Merlin OR 97532-9900

FOLD HERE SECOND, THEN TAPE TOGETHER

Please cut along this vertical line, fold twice, tape together and mail.